# Springer Atmospheric Sciences

More information about this series at http://www.springer.com/series/10176

Dana Magdalena Micu • Alexandru Dumitrescu
Sorin Cheval • Marius-Victor Birsan

# Climate of the Romanian Carpathians

## Variability and Trends

Springer

Dana Magdalena Micu
Institute of Geography of the Romanian Academy
Bucharest, Romania

Alexandru Dumitrescu
Sorin Cheval
Marius-Victor Birsan
National Meteorological Administration
Bucharest, Romania

ISSN 2194-5217 ISSN 2194-5225 (electronic)
ISBN 978-3-319-02885-9 ISBN 978-3-319-02886-6 (eBook)
DOI 10.1007/978-3-319-02886-6
Springer Cham Heidelberg New York Dordrecht London

Library of Congress Control Number: 2014953841



Printed on acid-free paper

Springer is part of Springer Science+Business Media (www.springer.com)

# Foreword

Mountain regions are fragile spots in many aspects, and they have always represented a key environment for the well being of the human communities. Mountains provide a substantial range of ecosystem services, like protection, biodiversity, water storage and supply, food, and recreation. At the same time, various natural and anthropogenic threats have increasingly endangered the sustainable development of such areas, while conservation and management strategies have been reinforced consequently. As a result of natural factors and anthropogenic bias, the climate shapes actively the mountain landscape, both at large and finer scales, with considerable consequences on any activity.

The Carpathian Mountains or Carpathians stretch over seven European countries. Along the history, different political, social and economical changes have modified the natural background, but common features have been also preserved. There is a real need for high quality, homogeneous and consistent data bases to address environmental issues, with applications in many fields.

About one third of the Carpathians lie over the territory of Romania, and the geographical location and morphological characteristics lead to important climatic differences between various sub-units. Conducted by a dynamic and competent group of Romanian climatologists, this study is a thorough and attractive investigation into the climate of the Romanian Carpathians, focusing mainly on the observed characteristics and variability along 1961–2010, by means of weather station records from Meteo Romania (National Meteorological Administration), which were quality controlled and homogenized within the project Carpatclim (Climate of the Carpathian Region). The authors have adjusted and analyzed the input data according to the objectives expressed in the introductory chapter, and the outputs and results are outstanding. *The Climate of the Romanian Carpathians. Variability and Trends* is the first comprehensive climatological study covering the entire Romanian Carpathians. Such a complex synthesis addresses both fundamental science and applications, becoming a precious tool for students, large public, stakeholders and policy makers. It brings up-to-date science for climatologists and for all mountain practitioners exploiting climatic information and I am strongly confident in its short and long term value.

Bucharest, Romania
March 26, 2014

Norel Rîmbu

# Acknowledgments

The authors warmly acknowledge Distinguished Professor Emeritus of Geography **Roger Barry** (CIRES, University of Colorado, Boulder, CO, USA), and Professor **Mihai Dima** (Faculty of Physics, University of Bucharest, Romania) for their kind availability to review this book in a professional manner.

# Contents

# Chapter 1
# Introduction

Mountain regions are a key issue at global scale, their importance having been recognized at the United Nations Conference on Environment and Development (Rio de Janeiro, 1992) and further underlined by designating the year 2002 as International Year of Mountains (IYM). The global warming put critical challenges to mountain environment and related ecosystem services as mountain regions proved to be particularly vulnerable to climatic fluctuations. From many perspectives, the Carpathian Mountains have a decisive role for the Central and South Eastern Europe, shaping the landscape, or influencing the culture and human development. This book tackles mainly the climate characteristics and the observed trends in the Romanian Carpathians (1961–2010), but it also presents estimations regarding the climate projections through the next decades (2021–2050). It is designed to address the needs constantly expressed by a large variety of users, from students and mass media to policy makers, land planning actors, water managers, and economic operators.

After this introductory overview, the reader may spot the theoretical framework, which supported the structure and the development of the book, emphasizing the key position of mountain areas in environmental research, and reviewing the existent literature. It has to be mentioned that this book is the first comprehensive synthesis on the climate of the Romanian Carpathians.

Based on a rigorous selection of outputs from previous publications, the main characteristics of the studied area are thoroughly approached in Chap. 3. The Romanian Carpathians and their geographical divisions are placed within the European context, concise geomorphologic and geologic settings are described, and the most significant hydrology, vegetation, and soil features are presented.

Both the meteorological and ancillary data, and the methods used in this book are in fully concordance with the international practice and standards, and the spatial and temporal resolutions were used in a flexible way for responding in the best manner to the purposes of this work. Chapters 4 and 5 are dedicated to the ample description of the datasets, ground meteorological network and methods specific to various applications, such as homogenization, statistics, spatialisation,



D.M. Micu et al., *Climate of the Romanian Carpathians*,
Springer Atmospheric Sciences, DOI 10.1007/978-3-319-02886-6_1

and climate change analysis. The use of homogeneous data, the significant number of variables, and the gridding procedures contain a high level of originality, at least for the studied area.

The main factors shaping the climate of the Romanian Carpathians are treated in Chap. 6. The influence of the geographical location, topography, and regional atmospheric circulation are approached quantitatively, and carefully described for each variable.

The regional and temporal climatic patterns of the Romanian Carpathians during the present climate (observations and measurements), and for the future decades are consistently tackled in Chaps. 7, 8, 9 and 10. The analysis uses homogeneous datasets, satellite products, and outputs of Regional Climate Models, addressing mainly the average state of the climatic elements, but carefully exploring the extremes as well. The result is a comprehensive, thorough, and synthetic document on the climatic background and perspective of the area. The characteristics of the solar radiation, air temperature, precipitation, wind and snow are analyzed statistically, emphasizing the mean state, and the extreme situations, specific to each geographic division of the studied area.

The future projections of the temperature and precipitation are treated in the Chap. 10, based on the results of several European projects. It is revealed that a drier climate may be expected to set up over in the Romanian Carpathians, as a result of higher temperature during all the seasons and relatively stable precipitation amounts.

The book *Climate of the Romanian Carpathians. Variability and trends* comes to cover with high quality, relevant climatic information a territory of multiple scientific, economic, and natural interests. It has been based on sing the experience and the results of previous work developed by the authors themselves, but also the outputs of other scholars. Despite their national coverage, the outputs contain a definite regional, transnational and trans-disciplinary dimension, due to the direct impact of the Romanian Carpathian Chain on other environmental or societal factors like hydrology, atmospheric dynamics, and economy.

The significant technological and scientific progress in data acquisition, homogenization, and sharing was an essential support for our task, and we would like to express our gratitude to all those who believe they deserve it. We hope the reader will find this book useful, and a good starting point for future approaches.

# Chapter 2
# Theoretical Background

**Abstract** The chapter is organized in two sections. The first section overviews the main impulses, initiatives and achievements in global and European environmental research targeting mountain regions since the early 1970s until 2012 (the Rio +20 Conference). This section also highlights how the growing significance of the climate change issue in global environmental research agendas and of the potential climate change impacts in mountain environments, has promoted mountain regions to a next stage of importance, as they are currently widely recognized as "early indicators of climate change". The main achievements of the Mountain Research Initiative in promoting the research in the Carpathian region and building opportunities for future collaborations targeting this region are also emphasized. The second section outlines the main contributions to the knowledge of the climate and weather of the Romanian Carpathians. The survey of the Romanian specialist literature shows a general lack of comprehensive studies carried out on the Romanian Carpathians as a whole, only a few aiming at revealing the regional patterns of a distinct climatic feature, usually derived from short period of observations of less than 10 years.

## 2.1 Mountain Regions: Key Issues in Environmental Research

Joint scientific-political initiatives on mountain environmental research date back to the early 1970s. One of the first initiatives of global environmental research in mountain regions was the Project 6 of the United Nations Educational, Scientific and Cultural Organization (UNESCO) on Man and the Biosphere (MAB), focusing on the "Impact of Human activities on Mountain Ecosystems". This project proved to bridge the natural and social sciences by addressing four key-impacts on mountain environments (land use alternatives, large-scale technology, human settlements, tourism and recreation activities) and also, to contribute to the development of methods, transdisciplinary approaches and collaborations (UNESCO 1973). The final declaration of the UN Conference on Human Environments (June 1972, Stockholm) provided also a great stimulus for subsequent



D.M. Micu et al., *Climate of the Romanian Carpathians*,
Springer Atmospheric Sciences, DOI 10.1007/978-3-319-02886-6_2

mountain research initiatives, even if mountain issues have been addresses indirectly. These early global impulses strengthened the opportunities during the 1980s, the 1990s and at the beginning of the twenty-first century, for setting up organizations, cooperation initiatives and conventions targeting different mountain regions, most of them still operational today e.g. the International Centre for Integrated Mountain Development (1983), the African (1986) and Andean (1991) Mountain Associations, Consortium for Sustainable Development of the Andean EcoRegion (1992), the Alpine (1991) and Carpathian (2001) Conventions.

Mountain research got a decisive impulse in 1992 at Rio de Janeiro, during the UN Conference on Environment and Development ("the Rio Earth Summit"), as a follow up of the 1972 UN Conference in Stockholm. The world's mountain regions gained significance as they were considered a priority altogether with other global change issues (e.g. climate change, desertification, deforestation). The awareness on the growing impact of human activities on mountain environments has increased significantly, further encouraging the interactions between science and politics. Recognising and emphasizing the global importance of mountain ecosystem services and goods, the Agenda 21 addressed the overall sensitivity of mountain ecosystems within its 13th Chapter, entitled: "Managing Fragile Ecosystems: Sustainable Mountain Development" (UN 1992). It is worth mentioning that the UNESCO recognized the importance of mountain regions several years before the 1992 Rio Summit, by promoting the contribution of these regions in several global programmes (MAB, United Nations University, World Climate Research Programme, International Geosphere-Biosphere Programme and DIVERSITAS: International Programme on Biodiversity Science) or specific programmes devoted to mountains (e.g. World Glacier Monitoring Service).

A wide range of efforts have been undertaken to promote mountain regions for creating opportunities for regional and global cooperation. On November 10th, 1998, the UN General Assembly declared the year 2002 the "International Year of Mountains" (IYM), as a living proof of the importance of mountain regions for the mankind. The IYM concept emphasized the activities related to the conservation and sustainable development of mountain regions. Its objectives aimed at promoting the interdisciplinary and integrated research on mountain ecosystems, the cultural heritage of mountain communities and peace initiatives in such areas, under international cooperation of neighbouring countries. The IYM proclaimed the day of 11th of December as the "International Mountain Day" (IMD), which has been celebrated since 2003. Proposing different themes yearly, the IMD aims to increase awareness on mountain regions issues and to develop partnerships for the sustainable development of these regions.

Later, during the World Summit on Sustainable Development held in Johannesburg (South Africa) in 2002, the concepts of IYM have been strengthened. In response, the Food and Agriculture Organisation (FAO) established in 2002 the International Partnership for a Sustainable Mountain Development. This partnership was designed to promote and support the national and international research initiatives on mountain regions, to strengthen building capacity and to implement strategies for the sustainable development of mountain ecosystems. These objectives were targeting particularly developing countries and those with a growing economy.

After 1992, mountain regions have received growing attention from three international global environmental change organisations – the International Geosphere-Biosphere Programme (IGBP), the International Human Dimensions Programme on Global Environmental Change (IHDP) and the Global Terrestrial Observing System (GTOS). The joint forces of the IHDP, IGBP and GTOS programmes created the basis of a new research programme – the Mountain Research Initiative (MRI), funded by the Swiss National Science Foundation (SNSF). The objectives, approaches and research activities of the MRI were altogether defined in 2001, recognising that 'mountain regions may experience the impacts of the rapidly changing global environment more strongly than others' (Becker and Bugmann 2001; Dexler 2008). This initiative aimed at promoting and coordinating global change research in the mountain regions worldwide, focusing on the monitoring of environmental change in mountain environments, investigation of the consequences of these changes for both mountain and lowland regions, as well as on the sustainable management of land-use and natural mountain resources at local-to-regional scales. The MRI along with other specific global programmes targeting mountain regions worldwide such as the Global Observation research Initiative in Alpine Environments (GLORIA) and the Global Mountain Biodiversity Assessment (GMBA), stand as evidence of the continuation of efforts for further promotion of research initiatives.

MRI has produced in 2005 a valuable science-oriented compendium entitled "Global change in mountain regions – An overview of current knowledge", which provided an overview of research contributions focusing on the detection, understanding and prediction of global change impacts in worldwide mountain regions (Huber et al. 2005). This publication created the basis for a FP6 cooperation within the GLOCHAMORE (GLObal CHAnge in MOuntain REgions) Project, with a further support of the UNESCO MAB programme, which aimed at elaborating an integrated and implemental research strategy to improve the understanding of causes and effects of global change in mountain regions worldwide. The MRI core activities are currently oriented to promote and implement the GLOCHAMORE strategy, through the initiation and support of regional networks of global change researchers.

Networking on mountain issues advanced significantly after the Rio Earth Summit in 1992, providing a solid base for the future European research activities (Dax 2002): Mountain Forum (1996), Euromontana (1996), European Mountain Forum (1998), the Alpine Convention (Convention on the Protection of the Alps) (2003), Charter for the Protection of the Pyrenees (1995), the Carpathian Ecoregion Initiative (1998), the Carpathian Convention (2003), Forum Carpathicum (2010).

With the strong support of the scientific community, the Conference on "Global Change and the World's Mountains", held on October 2005, at Perth (UK), promoted the necessity for further investigations of global change effects on mountain ecosystems, recognising that "global change, and in particular global warming, has and will have serious impacts on policies, the biophysical environment, and the socio-economic conditions and livelihoods of people, particularly in fragile mountain environments, but also in the adjacent lowland areas" (Perth Declaration 2005).

The Millennium Ecosystem Assessment is another major global research effort initiated in the framework of IYM. Its synthesis report entitled "Ecosystems and Human Well-being" (Millennium Ecosystem Assessment 2005) was focused on the consequences of ecosystem change for human well-being. Mountain ecosystems are among the ten systems being evaluated within this assessment (chapter 24 "Mountain systems").

### *2.1.1 Mountain Regions and Climate Change*

Documenting the main steps in the history of global mountain research, Messerli (2012) outlined that the growing significance of "climate change" for mountain water resources and of "biodiversity" and "ecosystem services" for both mountain and lowland populations. Schröter et al. (2005) emphasized the role of the scientific community in improving the awareness on the impacts of global environmental changes and climate change in particular, on the natural mountain ecosystems and their capacity to provide goods and services to the living in these regions and to the population of lowlands. Recent statistics have showed that 12 % of the World's population depends directly on mountain resources, while an even higher percentage uses precious mountain resources like water (FAO 2011).

Under the 2007 UN General Assembly Resolution 62/196 "the role of the scientific community, national governments and inter-governmental organisation is to collaborate with mountain communities in joint studies and address the negative effects of global climate change on mountain environments". Attempting to outline the need for coordinated adaptation strategies towards the expected climate change in the highly sensitive mountain environments of the world, the UNEP organised in 2008 at Padua (Italy) a Conference on "Mountains as Early Indicators for Climate Change". The conference goal was to provide support and to promote the exchange of up-to-date scientific research results concerning the changing climate signals and their effects in mountain regions worldwide. It was widely recognized that mountain regions are placed among the primary areas vulnerable to changing climate conditions.

Mountain ecosystems are emerging as vulnerable to climate changes (Beniston 1994; Sonesson and Messerli 2002). These regions were exposed to above-average warming during the twentieth century (IPCC 2007) and they are expected to be further exposed to climate warming and associated extremes over the twenty-first century. Nogues-Bravo et al. (2007) indicate that the average warming projected in mountain areas across the globe by 2055 is expected to range between 2.1 and 3.2 °C. Furthermore, Körner (2009) outlined that mountain regions are valuable early indicators of change in biodiversity.

Recognising the importance of mountain areas for the environment, society and economy of the European continent, where 36 % of its surface is mountainous, the European Environment Agency (EEA) outlined the vulnerability of mountain ecosystems to threats related to the land abandonment, intensifying agriculture, impacts of infrastructure development, unsustainable exploitation and last but not least, climate change and the related extremes (EEA 2010). EEA elaborated several reports (e.g. EEA 2008, 2009, 2010) addressing the environmental and climate change issues in various European mountain regions, including the Carpathians, which are considered a unique bio-geographic region (EEA 2008) of the continent. The overall findings of these reports suggested that the European mountain regions experienced severe climate change effects over the last decades, mostly expressed by glacier retreat, notable temperature increases, changes in precipitation regime and distribution.

The AR4 IPCC report (Trenberth et al. 2007) gave emphasis to mountain regions, reporting that these regions are particularly vulnerable to climate changes. IPCC outlined the importance of mountain glaciers which proved to be highly sensitive to temperature and precipitation oscillations, by considering them as the best terrestrial indicators of climate change.

The recognition of mountain regions among the highly vulnerable regions of the globe to climate change effects is also depicted in a distinct paragraph (94) within the Rio +20 Draft Agenda ("The future we want", January 10, 2012), which states: "We recognise that mountains are highly vulnerable to global changes such as climate change, and are often home to some communities including indigenous peoples, who although have developed sustainable uses of their resources, yet are often marginalised, sometimes with high poverty rates, exposure to natural risks and food insecurity. We recognise the benefits derived from mountains and their associated ecosystems. We also recognise the need to explore global, regional, national, and local mechanisms to compensate and reward mountain communities for the services they provide through ecosystem protection".

### *2.1.2 Mountain Research in the Carpathian Region*

In response to the research priorities defined within the GLOCHAMORE research strategy, a MRI Europe network was established in 2007, for supporting, promoting and coordinating the research activities in European mountain regions. The MRI Europe endorsed in 2008 the SC4 network (Science for the Carpathians), with the purpose of connecting scientists and their research from the Carpathian countries, identifying key issues for the Carpathian mountain research and also, building opportunities for future collaborations with partners from outside the Carpathians. The launch of the S4C initiative created the base for the development of a comprehensive "Carpathians Research Agenda (2010–2015): Integrating nature and society towards sustainability" (Ostapowicz and Sitko 2009; Kozak et al. 2011). The departure point of this agenda resulted from the GLOCHAMORE strategy

(Björnsen Gurung 2006). The Carpathians gained substantial visibility after the launch of the Forum Carpathicum (2010, Kraków), an interdisciplinary conference devoted to the Carpathian region, with the purpose of identifying the status and emerging issues in the current and future research, which bring together scientists, practitioners and stakeholders.

In April 2009, during the International Conference on "Identifying the Research Basis for Sustainable Development of the Mountain Regions in Southeastern Europe", held in Borovets (Bulgaria), it was launched the South Eastern European Mountain Research Network (SEEmore). The SEEmore goal was to provide support for scientific networking, research coordination and collaboration in mountain regions of the South-Eastern Europe. Since commissioned, the SEEmore confronts challenges associated to the rapid transformation in land use, biodiversity and bio-productivity and tourism due to the global changes and particularly due to the local impacts of climate change.

European mountain regions were generally under the focus of large research projects, funded mainly within the European framework (FP) and Interreg programmes. The climate change impacts on European mountain regions proved to be a target topic of several completed or ongoing research projects: e.g. FP4 PACE (Permafrost and climate in Europe: climate change, mountain permafrost degradation and geotechnical hazard), 1997–2001; GLORIA (Global Observation Research Initiative in Alpine Environments), 2004–2011; FP6 CLAVIER (Climate Change and Variability: Impact on Central and Eastern Europe), 2006–2009; Interreg III CLIMCHALP (Climate Change, Impacts and Adaptation Strategies in the Alpine Space), 2006–2008; FP6 CIRCLE (Climate Impact Research Co-ordination for a Larger Europe), 2005–2009; FP7 ACQWA (Assessment of climatic change and impacts on the quantity and quality of water), 2008–2013; FP7 CIRCLE-2 (Climate Impact Research & Response Coordination for a Larger Europe – 2nd Generation ERA-Net – Science meets Policy) with CIRCLE-2 MOUNTAIN, 2010–2014 etc. It is worth mentioning that, the Carpathians have been rarely the target region in projects focusing on environmental or climate change issues. Other relevant EU projects tackling the climate variability and climate change impacts in the Carpathian Mountains region were: the TATREX Polish-Czechoslovakian Programme (1981–1982), Climate Changes and Variability in the Western Carpathians project, CARPIVIA (2010–2013) and CarpathCC (2010–2013).

The importance of the Carpathian region at European scale has been widely recognized since 2003, with the launch of the Carpathian Convention. Later, in 2007, the outputs of a bottom-up collaborative and consultative project were officially released, with the support of the UNEP's Division of Early Warning and Assessment (DEWA)/GRID-Geneva and the Regional Office for Europe (ROE), under the form of a comprehensive environmental assessment and future outlook addressing the entire Carpathian Mountains region (Carpathians Environmental 2007). This synthesis put together the results obtained throughout national climate research projects from seven European countries (the Czech Republic,

Hungary, Poland, Romania, Serbia and Montenegro, the Slovak Republic and Ukraine).

Reviewing the status of global change research in the Carpathians, Björnsen Gurung et al. (2009) identified the main gaps in the main research fields such as climatology, hydrology, land use and land cover change, forestry, biodiversity and conservation, tourism and ecosystem services. The authors emphasized the need of collaboration actions targeting the Carpathian Region as a whole. The recommendations to overcome the gaps identified in the field of climatological research were to "establish of a joint international climatological database of long-term data, to set up additional meteorological stations in high-elevation areas and to make data freely available for scientists".

In response to the need of a pan-Carpathian research initiative, the JRC Carpatclim project (2010–2013) is an example of effort meant to compile and share climate data on a joint platform, overcame the lack of a high-quality and unitary database for the entire Carpathian Mountains region, aiming also at providing comprehensive scientific information for future climatological studies in the region. The project produced and released on June 2013 an open-access digital climate atlas of the Carpathian Mountains region, available with a $0.1 \times 0.1°$ horizontal resolution. The project outputs were based on homogenized meteorological datasets from ten European countries (including Romania), covering the 1961–2010 period, using a standard methodology.

## 2.2 Weather and Climate of the Romanian Carpathians: A Literature Review

Scientific climatology has emerged once visual and instrumental observations were being used in line with the international methodology. The first Romanian climatologist considered to have made significant contributions to this science is Ştefan C. Hepites. The data yielded by instrumental measurements at weather stations, processed and published later were used to elaborate several climate works of localities, regions or of the whole Romanian territory.

The first climatological works signed Ştefan C. Hepites was published in the *Annals of the Romanian Academy* (*Analele Academiei Române* in Rom.) between 1895 and 1902, and republished under the title *Materials devoted to Romania's climatology* (*Materiale pentru climatologia României*, in Rom.), for example, *The climate of Sinaia Town* (Hepites 1896) (*Clima Sinaiei*, in Rom.), one of the first mountain climatology studies in this country. The author makes a series of considerations on the characteristics of the main climatic variables (i.e. air temperature, relative atmospheric humidity, atmospheric precipitation, atmospheric pressure, wind and nebulosity), briefly outlining the days when various meteorological phenomena were recorded (e.g. thunderstorms associated with hail or snow hail, fog, snowfalls, snow days) and included in the Miscellanea chapter. The study is based on the results of measurements performed at Sinaia Monastery station (879 m a.s.l.), between 1886 (when the station was set up) and 1895.

Other studies on the climate features of the mountain region, published between the 1921 and 1930 interval are due to Enric Otetelişanu. The author discusses the influence of pressure centers in various parts of Europe and their effect on the climate of all of Romania's regions, illustrating his considerations with temperature and precipitation data, and analyzing the effects of geographical factors on the country's climate, including the Carpathian Mountains. Having in view that the mountain relief covers one-third of Romania's territory, the meteorological observations conducted at different elevations did contribute, yet much later (since 1925), to emphasizing the influence of the Carpathian Chain on the climate of the whole country. In 1929, Constantin A. Dissescu elaborated two studies with focus on air frost phenomena in the alpine areas of the Carpathians Dissescu (1929a) and on temperature variation with elevation in the Bucegi Massif Dissescu (1929b), both published in the Monthly Meteorological Bulletins of the Central Meteorological Institute.

The first and most comprehensive Romanian mountain climate monograph was elaborated in 1951 by Ştefan M. Stoenescu, focusing on the Bucegi Mountains (the Southern Carpathians).

The implementation of the new measurement scheduled in 1960 explains the progresses made later in the area of scientific climate research.

Some synthesis works published during the 1960s and highlighting the general features of Romania's climate included referees to the Romanian Carpathian region: e.g. *The climate of Romania*, 1961–1962 (*Clima României*, 1961–1962, in Rom.), the first synthesis of the country's climate; Topor (1963); *The climatological atlas of the Romanian People's Republic*, 1966 (*Atlasul climatologic al R.S.R.*, 1966, in Rom.).

The scientific works published in the periodicals of the Meteorological Institute (later Institute of Meteorology and Hydrology and currently the National Meteorological Administration) after 1960 stand proof to a fruitful scientific activity. Other researchers were dealing with some aspects, still topical today, such as the long-term variations of key climatic variables subject to change (air temperature and precipitation). Some of the remarkable works elaborated by Ştefan M. Stoenescu focused on climate oscillations and climate change in Romania (Stoenescu 1959, 1964), were actually the first approaches to these issues in this country. The endeavours of that period also had in view extreme phenomena, such as snowstorms (the severest twentieth-century event occurring in February, 1954), heavy rainfalls, snow avalanches, rime, sleet and glaze.

Climate research has been progressing in line with practical interests. The climatologists started to give more emphasis to the major influence of the underlying terrain on the characteristics of physical processes and boundary layer conditions (Topoglimatologia României. Bibliografie selectivă adnotată 1987). Interests in this area succeeded in advancing a new research direction in applied climatology both worldwide and in Romania, namely, topoclimatology. The first works on this topic were devoted to establishing the terminology specific to this new research direction in Romania. Vintilă Mihăilescu was the first in Romania to propose the term *topoclimate*, in a paper published in Russian (Mihăilescu 1957), defining it as "the climate of the contact zone of planetary covers considered over small areas, thus making it possible to analyze at local level the relationships between the physical phenomena in the atmosphere and the other components of the geographical complex or environment (firstly, landform and secondly, water, vegetation, etc.)". Later, together with Ştefan C. Stoenescu, elaborated the second edition of the *map of climate and topoclimate of Romania* (Mihăilescu and Stoenescu 1960), bringing concise data and quantitative indexes to sustain each individualized topoclimate. This map on the scale of 1:3,000,000 depicts the climatic potential of several topoclimatic units, among which the mountainous topoclimatic belt with its sub-belts: alpine, subalpine and the mountain proper.

Several contributions to this new research direction emphasized some aspects related to the topoclimatic features of some cities, spas and tourist resorts (e.g. Bogdan and Mihai 1977; Teodoreanu 1981, 1997; Bogdan and Niculescu 1996a) and the role of the subjacent topographic surface as a topoclimate generator (e.g. Niculescu 1993; Bogdan and Niculescu 1996b).

The Romanian Carpathians as a whole, or their three component units, have fairly seldom made the object of complex mountain climatology studies (syntheses). Regional climate studies, targeting smaller geographical areas within the Romanian Carpathians, were started mainly after 1960 and focused preferentially on certain climatic elements, or on the natural factors influencing them:

- **Air temperature regime** – e.g. Dissescu (1929a, b) and Şorodoc (1960), temperature distribution with altitude in the Bucegi Mountains; Stoenescu and Dumitrescu (1965), the frequency and daily average temperature in the Bârsa Depression; Dumitrescu et al. (1971), the frequency of positive minimum winter temperature in the Giurgeu, Ciuc, Bârsa, Făgăraş and Sibiu depressions; Bogdan and Mihai (1972), thermal amplitude in the Romanian Carpathians; Neacşa et al. (1972), the variation of thermal parameters in relation to topographic mountain features; Dobrea (1972), air temperature characteristics in the mountain sector of the Prahova Valley; Oprescu and Bogoriţă (1976), air temperature distribution in the Lotru-Parâng Mountains (Southern Carpathians); Măhăra and Linc (1993), thermal anomalies in the Codru-Moma Mountains (Western Carpathians); Dragotă and Gaceu (2005), extreme temperatures in the Bihor and Vlădeasa Mountains (Western Carpathians); Gaceu (2004), air temperature distribution in the Bihor and Vlădeasa Mountains; Popa and Cheval (2007), early winter temperature recontruction of Sinaia area (Southern Carpathians) derived from tree-rings of silver fir; Cheval et al. (2011), July surface temperature gradient in the Romanian Carpathians, Croitoru et al. (2011), change-point analysis if temperature variability at high elevation sites etc.;
- **Precipitation regime** – e.g. Marcu (1967), the precipitation regime in the Postăvaru Massif and Bârsa Depression; Teodoreanu (1972), the frequency of 1-day maximum precipitation in the Southern Carpathians; Neamu and

Teodoreanu (1972), altitudinal distribution of precipitation amounts in the Romanian Carpathians; Teodoreanu (1973), the annual precipitation regime in the Rucăr-Bran Corridor (Southern Carpathians); Pătăchie et al. (1976), distribution of precipitation in the Parâng Massif (Southern Carpathians); Popa and Dragotă (1978), the precipitation regime in the Romanian Carpathians; Stăncescu et al. (1986), the particularities of precipitation distribution induced by frontal process in the Romanian Carpathians and its surrounding areas; Moldovan and Bretan (1989), precipitation regime at high altitude stations (Vlădeasa 1,400 m and Vlădeasa 1,800 m) in the Vlădeasa Mountains (Western Carpathians); Zăvoianu et al. (1995), altitudinal distribution of precipitation in the Southern Carpathians; Măhăra and Gaceu (2005a), a study of heavy rainfalls in the western part of the Apuseni Mountains (Western Carpathians); Gaceu and Linc (2005), the precipitation regime in the Bihor and Vlădeasa Mountains; Mic et al. (2008), the scale and shape variables of the precipitation regime in the Bucegi Mountains and the Prahova Corridor (Southern Carpathians), etc.;

- **Wind regime** – e.g. Constantin and Cristescu (1967), wind regime in the Țarcu Massif (Southern Carpathians); Șerbu (1968), local air flows in the Semenic Mountains (Western Carpathians); Dobrea and Patrichi (1969), wind regime in the Bucegi Mountains and surroundings; Belozerov et al. (1984), aeolian potential in the North-Western sector of the Vlădeasa Mountains; Pitea et al. (1986), atmospheric pressure oscillations during strong winds at Vlădeasa 1,800 m weather station; Bogdan (1985), the characteristics of atmospheric calm in depression areas and mountain valleys; Măhăra and Gaceu (2005b), strong winds and wind gusts on the western side of the Apuseni Mountains; Dragotă et al. (2005), wind regime in the Bihor and Vlădeasa Mountains, etc.;
- **Snowfalls and snow cover** – e.g. Topor (1951), snow onset in Romania; Țâștea (1959), snowpack stability; Marcu (1970), snow regime in the Postăvaru Massif (Eastern Carpathians); Gugiuman and Stoian (1972), snowfall and snow cover regime of the Romanian Carpathians (1961–1969); Raț (1973), on the snowmelting season in Romania; Voiculescu (1995), snowfall and snow cover regime at Bâlea Lake weather station (Southern Carpathians); Bogdan and Iliescu (1999), comparative climatic and topoclimatic features of snow cover in the Apuseni and Bucegi Mountains; Geicu et al. (1992), snow stability in Romania; Dragne et al. (2004), snow cover and the factors influencing it in the Romanian Carpathians; Flueraru et al. (2004) and Flueraru et al. (2005), the characterisation of snowpack key parameters derived from remote sensing; Micu and Mic (2006), snowfall regime in the Romanian Carpathians; Iosub (2007), spatio-temporal distribution of snow cover areas in the Bucegi Mountains; Micu (2009a), snowpack in the Romanian Carpathians under changing climatic conditions; Micu and Mărășoiu (2009), heavy snowfalls in the Southern Carpathians; Micu (2009b), the frequency of rapid snowmelt episodes in the Romanian Carpathians, etc.

After 1990, climate variability and change and weather extreme events, became topics of a major concern for many researchers in Romania. Several climatological syntheses (at national or regional scales) made references to the Carpathian region and are worth mentioning:

- **Climate variability and change**: e.g. Bălteanu et al. (1987), climate change impact on the mountain regions of Central Europe; Voiculescu (2002), climate warming effects on the subalpine areas of the Southern Carpathians; Micu and Micu (2006), winter temperature variability in the Romanian Carpathians; Micu (2007), variability of air freeze regime in the Romanian Carpathians; Bojariu and Dinu (2007), snow cover variability and trends in the Romanian Carpathians; Micu (2008), precipitation variability in the Southern Carpathians; Micu and Dincă (2008), winter climate change and its effects on winter tourism in the Prahova Valley-Poiana Braşov mountain area; Micu (2009a), snowpack changes under changing climatic conditions; Micu and Mic (2009), variability and trends in the snow regime parameters of the Romanian Carpathians; Busuioc et al. (2010), climate variability and change in Romania; Croitoru et al. (2011), change-point years; Cheval et al. (2014) and Spinoni et al. (2014), climate variability in the Carpathian Region.
- **The link between climate change and atmospheric circulation**: e.g. Busuioc and von Storch (1996), winter precipitation change and its link to large-scale circulation; Tomozeiu et al. (2002), seasonal change of maximum temperature in Romania and the influence of large-scale circulation;
- **Weather extreme events**: e.g. Hauer et al. (2003), trends in temperature and precipitation extremes in the north-western part of Romania (including two mountain weather stations located above 1,700–1,800 m); Bogdan (1998), absolute temperature records in Romania; Croitoru et al. (2002), effects of extreme temperatures and the 1-day maximum precipitation in the north-western part of Romania; Busuioc et al. (2003), variability of precipitation extremes in Romania; Cheval et al. (2004), variability of climate extremes in the Romanian Carpathians; Bogdan et al. (2007), warm winters in Romania (with focus on the 2006/2007 warm winter); Birsan et al. (2014a), changes in the temperature extremes in the Carpathians;

Snow avalanches and their triggering factors in the Carpathian region are topics addressed both by old (e.g. Stoenescu 1956; Drâmbă 1957) and especially by several recent studies, elaborated after 2004, when the National Nivo-Meteorology Programme of the National Meteorology Administration became operational. Most of these studies focused on the highest and steepest massif of the Romanian Carpathians, with well-developed glacial landforms, namely the Făgăraş Mountains (the Southern Carpathians): e.g. Moţoiu et al. (2004, 2005), Voiculescu et al. (2004a, b, 2007, 2009, 2010, 2012), Moţoiu (2008), Câmpean and Câmpean (2010), Voiculescu (2014), Voiculescu and Onaca (2014).

Several remote sensing and hydrologic studies targeted different areas of the Romanian Carpathians, made contributions to the snowpack research, by considering it a valuable water resource, but also a predictor of the run-off variation: e.g. (Moissiu 1980, 1983; Grumăzescu et al. 1986; Miţă and Drăgan 1986; Grumăzescu and Stăncălie 1988; Copaciu and Tibacu 1993; Diaconu and Şerban

1994; Miță and Stăncălie 1996; Miță et al. 2003; Zăvoianu et al. 1999; Stăncălie et al. 2004; Birsan et al. 2012, 2014b).

Relevant contributions to the Romanian mountain research have been made under the framework of several projects, generally addressing topics related to climate change and high-mountain natural risks:

- Post-glacial dynamics of alpine environments in the Romanian Carpathians. Morphoclimatic correlations, geotopes preservation and sustainable landscape management (MEDALP project, 2006–2009);
- GIS-related elaboration of database and thematic maps of ski areas in the Southern Carpathians. Analysis, assessment and prognosis in the light of global climate change (2009–2011);
- Natural risk phenomena in the Southern Carpathian alpine realm. GIS-based risk mapping (2004–2007).

## References

Atlasul climatologic al R.S.R (1966) The climatological atlas of Romania. CSA, IM, București (in Romanian)

Bălteanu D, Ozenda P, Kuhn M, Kerschner Tranquillini W, Bortenschlager S (1987) Impact analysis of climatic change in the Central European mountain ranges. In: Proceedings of the European workshop on interrelated bioclimatic and land changes. Noordwij Kerkout, pp 6–41

Becker A, Bugmann H (eds) (2001) Global change and mountain regions: the mountain research initiative. IHDP Report 13, GTOS Report 28 and IGBP Report 49, Stockholm

Belozerov V, Buta I, Sorocovschi V, Maier A, Schreiber W, Tudoran P, Fărcaș I, Idu P (1984) Aspecte privind evaluarea potențialului eolian în sectorul nord-verstic al Masivului Vlădeasa, cu privire specială la stația Râșnov-Floroiu (Considerations on wind potential in the north-western sector of the Vlădeasa Massif, with focus on the Râșnov-Floroiu station). SUBB-GG XXIX:37–45 (in Romanian)

Beniston M (ed) (1994) Mountain environments in changing climates. Routledge, London, p 461

Birsan MV, Zaharia L, Chendes V, Branescu E (2012) Recent trends in streamflow in Romania (1976–2005). Rom Rep Phys 64(1):275–280

Birsan MV, Dumitrescu A, Micu DM, Cheval S (2014a) Changes in annual temperature extremes in the Carpathians since AD 1961. Nat Hazards. doi:10.1007/s11069-014-1290-5

Birsan MV, Zaharia L, Chendes V, Branescu E (2014b) Seasonal trends in Romanian streamflow. Hydrol Process 28(15):4496–4505. doi:10.1002/hyp.9961

Björnsen Gurung A (ed) (2006) Global change and mountain regions research strategy. MRI/UNESCO Publication, ADAG Copy AG, Zürich

Björnsen Gurung A, Bokwa A, Chełmicki W, Elbakidze M, Hirschmugl M, Hostert P, Ibisch P, Kozak J, Kuemmerle T, Matei E, Ostapowicz K, Pociask-Karteczka J, Schmidt L, van der Linden S, Zebisch M (2009) Global change research in the Carpathian mountain region. Mt Res Dev 29(3):282–288

Bogdan O (1985) Particularități ale calmului atmosferic din unele regiuni depresionare și culoare de vale (Particularities of atmospheric calm in some depression areas and valley corridors. SCGGG-Geogr XXXII:14–19 (in Romanian)

Bogdan O (1998) Aspecte de risc în repartiția temperaturii extreme absolute în România (Risk aspects related to the absolute extreme temperature in Romania). Acad Șt Econom 1–2. București. pp 29–31 (in Romanian)

Bogdan O, Iliescu M (1999) Aspecte climatice şi topoclimatice comparative privind stratul de zăpadă din Munţii Bihor şi Munţii Bucegi (Comparative climatic and topoclimatic aspects on the snowpack in the Bihor and Bucegi Mountains). Rev Geogr V 1998. Serie nouă: 59–67 (in Romanian)

Bogdan O, Mihai E (1972) Amplitudinile de temperatură pe teritoriul Carpaţilor Româneşti (Temperature amplitudes across the Romanian Carpathians). Lucr. Simpoz. Geogr. Fiz. Carpaţi, Inst. Geogr., pp 203–212 (in Romanian)

Bogdan O, Mihai E (1977) Particularităţi climatice şi topoclimatice ale oraşului Sinaiaşi împrejurimilor sale (Climatic and topoclimatic particularities of Sinaia Town and its surroundings). Stud Cercet Geogr şi Geol Geogr XXIV(2):213–226 (in Romanian)

Bogdan O, Niculescu E (1996a) Valorificarea turistică a potenţialului climatic din Culoarul Prahovei şi Masivul Bucegi (Touristic use of the climatic potential of the Prahova Corridor and Bucegi Mountains). Anal Şc Prah Geogr II:26–42 (in Romanian)

Bogdan O, Niculescu E (1996b) Caractérististiques climatiques et topoclimatiques des certains culoirs des Carpates Méridionales. Geogr Int Sem 3. Bucureşti, pp 50–55

Bogdan O, Marinică I, Rusan N, Rusu S (2007) Riscul iernilor calde în România – cu aplicaţii la iarna 2006–2007 (The risk of warm winters in Romanian – the case study of the 2006–2007 Winter). Riscuri şi Catastrofe VI(4):97–110 (in Romanian)

Bojariu R, Dinu M (2007) Snow variability and change in Romania. In: Proceedings of alpine*snow*workshop, Forschungsbericht (53):34–38

Busuioc A, von Storch H (1996) Changes in winter precipitation in Romania and its relation to the large scale circulation. Tellus 48A:538–552

Busuioc A, Bororeanţ C, Matei D (2003) Analiza variabilităţii frecvenţei evenimentelor pluviometrice extreme în România (Analysis of the variability of extreme precipitation events frequency in Romania). Anal Univ Ovidius-Geogr (1):43–50 (in Romanian)

Busuioc A, Caian M, Cheval S, Bojariu R, Boroneanţ C, Baciu M, Dumitrescu A (2010) Variabilitatea şi schimbarea climei în România (Variability and climate change in Romania). Edit. ProUniversitaria, Bucureşti, p 226 (in Romanian)

Câmpean ON, Câmpean I (2010) Riscul de avalanşă în Carpaţii Româneşti. Studiu de caz Munţii Făgăraş – sezonul 2008–2009 (The snow avalanche risk in the Romanian Carpathians. Case study the Făgăraş Mountains – the 2008–2009 season). Riscuri şi Catastr 8(2):103–111 (in Romanian)

Carpathians Environmental Outlook (2007) United Nations Environment Programme/Division of Early Warning and Assessment, p 232

Cheval S, Baciu M, Breza T (2004) The variability of extreme events in the Romanian Carpathians. Anal Univ de Vest din Timişoara-Geogr (XIV): 59–78

Cheval S, Al D, Petrişor A (2011) The July surface temperature lapse in the Romanian Carpathians. Carpath J Earth Environ Sci 6(1):189–198

Cheval S, Birsan MV, Dumitrescu A (2014) Climate variability in the Carpathian Mountains Region over 1961–2010. Global Planet Change 118:85–96. doi:10.1016/j.gloplacha.2014.04.005

Clima României (1961–1962) (The Climate of Romania). Institutul Meteorologic, Bucureşti (in Romanian)

Constantin T, Cristescu Ş (1967) Regimul vântului în Masivul Ţarcu (Wind regime in the Ţarcu Massif). Cul Lucr Meteor 1965, IM, pp 311–323 (in Romanian)

Copaciu V, Tibacu L (1993) Correlation between the density of the snow cover and meteorological parameters. In: Proceedings of the international symposium on precipitation and evaporation, Slovak Hydrometeorological Institute, Bratislava, pp 323–328

Croitoru A, Hauer E, Mihăilescu M (2002) Riscuri determinate de temperaturile extreme şi de cantităţile maxime de precipitaţii căzute în 24 ore în nord-vestul României (The risks associated with extreme temperatures and greatest 1-day precipitation in the northwestern Romania). Riscuri şi Catastrofe (I): 98–109

Croitoru A, Drignei D, Holobâcă I-H, Dragotă C (2011) Change-point analysis for serially correlated summit temperatures in the Romanian Carpathians. Theor Appl Climatol. doi:10.1007/s00704-011-0508-7
Dax T (2002) Research on mountain development in Europe: overview of issues and priorities. Available on http://lib.icimod.org/record/11031/files/22.pdf. Accessed 2 Apr 2014
Dexler C (2008) The mountain research initiative: from theory to practice. IHDP Updat 2:7–11
Diaconu C, Şerban P (1994) Sinteze şi regionalizări hidrologice (Hydrological synthesis and regionalizations). Editura Tehnică, Bucureşti (in Romanian)
Dissescu CA (1929a) Studiu asupra primei şi ultimei zile cu îngheţ în ţinuturile alpine (The study of the first and last air freezing day in alpine environments). Bul Meteor Lunar (October), IMC, Bucureşti (in Romanian)
Dissescu CA (1929b) Variaţia temperaturii cu altitudinea în Masivul Bucegi (Vertical variation of air temperature in the Bucegi Massif). Bul Meteor Lunar (August), IMC, Bucureşti (in Romanian)
Dobrea I (1972) Unele caracteristici ale regimului temperaturii aerului în sectorul montan al Văii Prahovei (Characteristics of air temperature regime in the upper Prahova Valley). Lucr Simpoz Geogr Fiz Carpat 1970: 240–251 (in Romanian)
Dobrea I, Patrichi S (1969) Caracteristicile locale ale regimului eolian în Masivul Bucegi şi zona înconjurătoare (Local features of wind regime in the Bucegi Massif and its surroundings). Cul. Lucr. Meteor./1967 (in Romanian)
Dragne D, Cheval S, Micu M (2004), The snow cover in the Romanian Carpathians and the influencing factors. In: Proceedings of the alpine geomorphology and mountain hazards workshop, Bâlea Cascadă (Romania), Sept 23–26
Dragotă C, Gaceu O (2005) Des considérations sur les températures extrêmes dans les montagnes de Bihor et Vlădeasa. Anal Univ din Oradea-Geogr (2): 84–88
Dragotă C, Gaceu O, Hârsescu S (2005) Regimul eolian din Munţii Bihor şi Vlădeasa (Wind regime in the Bihor and Vlădeasa Mountains). Anal Univ din Oradea-Geogr. (XIV): 85–92 (in Romanian)
Drâmbă D (1957) Fenomenul avalanşelor de zăpadă (Snow avalanches). Natura (5): 138–144 (in Romanian)
Dumitrescu E, Glaja M, Ciulache S (1971) Frecvenţa intervalelor cu temperaturi minime pozitive de iarnă în depresiunile Giurgeu, Ciuc, Bârsei, Făgăraş şi Sibiu (The frequency of the intervals with positive minimum temperatures during winter in the Giurgeu, Ciuc, Bârsa, Făgăraş and Sibiu depressions). Anal Univ din Bucureşti-Geogr (XX): 89–94 (in Romanian)
EEA (2008) Impacts of Europe's changing climate – 2008 indicator-based assessment. Report no. 4. Joint EEA-JRC-WHO report. p 19
EEA (2009) Regional climate change and adaptation The Alps facing the challenge of changing water resources. Report no. 8. European Environment Agency. p 143
EEA (2010) Europe's ecological backbone: recognising the true value of our mountains. Report no. 6. European Environment Agency. p 248
FAO (2011) Forests for improved nutrition and food security. Rome
Flueraru C, Stăncălie G, Savin E, Catană S (2004) Snow cover measurements in the Carpathian Mountains using satellite imagery, Analele Universităţii de Vest din Timişoara-Geogr. (XIV): 133–144
Flueraru C, Stăncălie G, Savin E, Catană S (2005) Determinarea acoperirii cu zăpadă în munţii Carpaţi folosind imagini satelitare de medie rezoluţie (Assessing snow cover in the Carpathian Mountains using satellite images of medium resolution). Rom J Climatol (I): 329–338 (in Romanian)
Gaceu O (2004) Caracteristici ale temperaturii aerului în Munţii Bihor şi Vlădeasa (Characteristics of air temperature in the Bihor and Vlădeasa Mountains). Comunicări de Geografie IX:189–197 (in Romanian)

Gaceu O, Linc R (2005) Aspecte privind regimul precipitaţiilor atmosferice în Munţii Bihor şi Vlădeasa (Considerations on precipitation regime in the Bihor and Vlădeasa Mountains). Rom J Climat 1:203–210 (in Romanian)

Geicu A, Popovici A, Chiriac D (1992) Consideraţii privind stabilitatea stratului de zăpadă (Considerations on snowpack stability). Studii şi Cercet-Meteor (6): 157–161 (in Romanian)

Grumăzescu H, Stăncălie G (1988) The surveillance of snow cover dynamics and the forecast of the snowmelt runoff in the Carpathian Basin using remote sensing data. In: Proceedings of the international workshop on hydrology of mountain areas, Strbske Pleso, Cehoslovacia

Grumăzescu H, Grumăzescu C, Stăncălie G, Săndoiu I (1986) Utilizarea datelor de teletecţie în modelele de topire a zăpezii – scurgere (Using teledetection in snowmelting-runoff models). Studii şi Cercet-Hidrol (53): 163–169 (in Romanian)

Gugiuman I, Stoian R (1972) Zăpada în Carpaţii Româneşti (Snowpack in the Romanian Carpathians). Lucr Simpoz Geogr Fiz Carpat 1970: 283–296 (in Romanian)

Hauer E, Croitoru A, Pavai C (2003), Tendinţele şi anomaliile extremelor termice şi pluviomterice în regiunea de nord-vest a României (Trends and deviations of extreme temperatures and precipitation in the northwestern Romania). Riscuri şi Catastrofe (I): 85–96 (in Romanian)

Hepites ŞC (1896) Clima Sinaiei/Le climat de Sinaia. Analele Institutului Meteorologic al Romaniei pe anul 1895. Tome XI(2):B165–B180 (in Romanian and French)

Huber UM, Bugmann HKM, Reasoner MA (eds) (2005) Global change and mountain regions: an overview of current knowledge. In: Advances in global change research, vol. 23, Springer, Dordrecht, 650 p

IPCC (2007) Climate Change 2007: synthesis report. Contribution of working groups I, II and III to the fourth assessment report of the intergovernmental panel on climate change. Core Writing Team, Pachauri RK and Reisinger A (eds) IPCC, Geneva, p 104

Iosub F (2007) Distribuţia spaţio-temporală a arealelor acoperite cu zăpadă în Munţii Bucegi (Spatial-temporal distribution of snow cover areas in the Bucegi Mountains). Online paper, available at http://earth.unibuc.ro/articole/distributie-spatio-temporala-zapada-bucegi. Accessed 25 Sept 2014

Körner C (2009) Conservation of mountain biodiversity in the context of climate change, In Proceedings of the International Mountain Biodiversity Conference, Kathmandu, 16–18 Nov 2008, ICIMOD, Kathmandu

Kozak J, Björnsen Gurung A, Ostapowicz K (eds) (2011) Research agenda for the Carpathians 2010–2015: integrating nature and society towards sustainability. Science for the Carpathians, Kraków, p 43

Măhăra G, Gaceu O (2005a) Les pluies torrentielles de l'ouest des Montagnes d'Apuseni (Les Carpates Occidentaux de la Roumanie et leur impacte sur l'environnement), Climat urbain, ville et arhitecture, Genes, pp 155–158

Măhăra G, Gaceu O (2005b) Fenomene de risc produse de vânturile tari şi vijelii în vestul Munţilor Apuseni (Climatic risks associated to strong winds and squalls in the Apuseni Mountains). Rom J Climatol 1:47–55 (in Romanian)

Măhăra G, Linc R (1993) Anomalii termice în Munţii Codru-Moma (Temperatures anomalies in the Codru-Moma Mountains). Anal Univ din Oradea-Geogr (III): 54–57 (in Romanian)

Marcu M (1967) Unele caracteristici ale regimului precipitaţiilor în Masivul Postăvaru şi în Depresiunea Bârsei (Some characteristics of precipitation regime in the Postăvaru Massif and Bârsa Depression). Bul. IP Braşov-B (IX) (in Romanian)

Marcu M (1970) Stratul de zăpadă în Masivul Postăvaru (Snowpack in the Postăvaru Massif). Bul IP Braşov-B (XII): 37–46 (in Romanian)

Messerli B (2012) Global change and the World's mountains. Mt Res Dev 32(S1):S55–S63

Mic L, Micu D, Dragotă C (2008) Parametrii de scară şi formă specifici regimului pluviometric în Masivul Bucegi şi Culoarul Prahovei (Scale and shape parameters of precipitation regime in the Bucegi Massif and Prahova Corridor). Comun de Geogr (XII) (in Romanian)

Micu D (2007) Variabilitatea îngheţului în aer în Carpaţii Româneşti (Variability of air freeze in the Romanian Carpathians). Comun de Geogr (XI): 18–25 (in Romanian)

Micu D (2008) Variabilitatea cantităților de precipitații în Carpații Meridionali (Precipitation variability in the Southern Carpathians). Rev Geogr (XIV–XV). Serie nouă: 18–26 (in Romanian)
Micu D (2009a) Snow pack in the Romanian Carpathians under changing climatic conditions. Meteorol Atmos Phys 105(1–2):1–16
Micu D (2009b) Frecvența topirilor rapide ale stratului de zăpadă în Carpații Românești (Frequency of rapid snowmelts in the Romanian Carpathians). Comun de Geogr (XIII): 121–133 (in Romanian)
Micu D, Dincă A (2008) Climate variability and its implications for winter mountain tourism in the Prahova Valley-Poiana Braşov area: proceedings of the international conference on ecological performance in a competitive economy, vol. I. Qual Access Success 9 (94): 232–241.
Micu D, Mărăşoiu D (2009) Ninsorile abundente din Carpații Meridionali (Heavy snowfalls in the Southern Carpathians). Revista Geografică (XVI). Serie nouă: 5–14 (in Romanian)
Micu D, Mic L (2006) Caracteristicile ninsorilor în Carpaţii Româneşti (Characteristics of snowfall regime in the Romanian Carpathians). Comunicări de Geografie X:165–173 (in Romanian)
Micu D, Mic L (2009), Trends and variability of the significant parameters of snow pack regime in the Romanian Carpathians. Riscuri şi catastrofe (6)
Micu D, Micu M (2006) Winter temperature trends in the Romanian Carpathians – a climate variability index. Anal Univ de Vest din Timişoara-Geogr (XVI): 141–159
Mihăilescu V (1957) O schiță de hartă topoclimatică a României (A sketch of topoclimatic map of Romania). Bul Şt. seria Geol.-Geogr II (3–4) (in Romanian)
Mihăilescu V, Stoenescu ŞM (1960) La carte climatiques et topoclimatique de la Roumanie. Recueil Etud Géogr: 55–64
Millennium Ecosystem Assessment (2005) Ecosystems and human well-being: synthesis. Island Press, Washington, DC., p 155
Miță P, Drăgan L (1986) Particularități privind scurgerea din topirea zăpezii (Particularities of snowmelting runoff). Studii şi Cercetări-Hidrol (53): 59–71 (in Romanian)
Miță P, Stăncălie G (1996) Methods of computing of the variation of the characteristics of the snow cover and runoff generated by snowmelting. Centre d'Etudes et des Recherches Eco-Géographique (CEREG), Strasbourg
Miță P, Simota M, Ungureanu V, Mătreață S (2003) Floods of combined origin – rainfall and snowmelt – in Romanian small catchments, proceedings of the 9th ERB conference, Slovakia, Sept 25–28, 2002, IHP-VI Technical Documents in Hydrology, p 67
Moissiu C (1980) Schemă de recunoaştere nivometrică într-un bazin hidrografic (A scheme of snowpack assessment in a catchment). Studii şi Cercetări-Hidrol. (XLVIII): 63–77 (in Romanian)
Moissiu C (1983) Teledetecția prin termoviziune aplicată în nivometrie (Thermal-based remote sensing applied in snowpack measurements). Studii şi Cercetări-Hidrol L(50): 61–71 (in Romanian)
Moldovan F, Bretan M (1989) Considerații asupra regimului precipitațiilor atmosferice de la două stații din Masivul Vlădeasa – Munții Apuseni (Considerations on the precipitation regime at two weather stations of the Vlădeasa Massif – Apuseni Mountains). Studii şi Cercetări-Meteor. (3): 283–291 (in Romanian)
Moțoiu M (2008) Avalanşele şi impactul lor asupra mediului. Studii de caz din Carpații Meridionali (Snow avalanches and their impact on environment. Case studies from the Southern Carpathians). Edit. Proxima, Bucureşti, p 279 (in Romanian)
Moțoiu M, Milian N, Truşcă M (2004) Fenomene extreme din Munții Făgăraş – Episodul de avalanşe din 01–07 martie 2004 (Valea Bâlea) (Extreme events in the Făgăraş Mountains – the snow avalanche of March 1–7, 2004 (the Bâlea Valley). Paper presented at the Annual Session of the National Meteorological Administration, Bucharest (in Romanian)
Moțoiu M, Milian N, Truşcă M, Flueraru C, Vlad T, Nica T, Tava A, Porceanu G (2005) Studii de caz – avalanşe provocatoare de pagube sau victime în Valea Capra, Munții Făgăraş, în sezonul

de iarnă 2004–2005 (Case studies – snow avalanches causing economic or human losses in the Capra Valley, the Făgăraş Mountains, during the 2004–2005 season). Paper presented at the Annual Session of the National Meteorological Administration, Bucharest (in Romanian)

Neacşa O, Popa G, Popovici C, Dincă Ileana (1972) Variaţia unor parametrii ai temperaturii aerului în Carpaţii Româneşti (Variation of some air temperature parameters in the Romanian Carpathians). Lucr Simpoz Geogr Fiz Carpat 1972: 221–229 (in Romanian)

Neamu G, Teodoreanu E (1972) Repartiţia precipitaţiilor atmosferice în raport cu altitudinea în Carpaţii Româneşti (Precipitation distribution by height in the Romanian Carpathians). Lucr Simpoz Geogr Fiz Carpat 1972: 275–283 (in Romanian)

Niculescu E (1993) Influenţa suprafeţei active montane româneşti asupra climatului şi topoclimatului de tip carpatic (Influence of the mountain underlying surface on the Carpathian climate and topoclimate). Anal Univ din Oradea-Geogr (III): 64–68 (in Romanian)

Nogués-Bravo D, Araújoc MB, Erread MP, Martínez-Ricad JP (2007) Exposure of global mountain systems to climate warming during the 21st Century. Glob Environ Chang 17:420–428

Oprescu A, Bogoriţă N (1976) Particularităţi ale repartiţiei temperaturii aerului în Munţii Lotru-Parâng (Particularities of air temperature distribution in the Lotru-Parâng Mountains). Studii şi Cercetări-Meteor I/2: 599–614 (in Romanian)

Ostapowicz K, Sitko I (eds) (2009) Science for the Carpathians strategy development and networking workshop, 27–28 May 2008. Report. Bern, Switzerland

Pătăchie I, Enăşescu G, Popa A (1976) Repartiţia precipitaţiilor atmosferice în Masivul Parâng (Precipitation distribution in the Parâng Mountains). Stud Cercetări-Meteor I/2:457–481, in Romanian

Perth Declaration (2005) Declaration on global change affecting mountain biosphere reserves. Available on http://www.unesco.org/mab/doc/mountains/PerthDecl.pdf. Accessed 1 Apr 2014

Pitea V, Vamoş C, Tilinca Z, Lungu V, Matei N (1986) Variaţia oscilaţiilor presiunii în cazul vânturilor tari la staţia Vlădeasa – Munţii Apuseni (Variation of atmospheric pressure during strong winds at Vlădeasa station – the Apuseni Mountains). Studii şi Cercetări-Meteor 1:113–117 (in Romanian)

Popa I, Cheval S (2007) Early winter temperature reconstruction of Sinaia area (Romania) derived from tree-rings of silver fir (Abies alba mill.). Rom J Meteorol 9(1–2):47–54

Popa A, Dragotă C (1978) Consideraţii asupra regimului precipitaţiilor atmosferice din Carpaţii Româneşti (Considerations on precipitation regime in the Romanian Carpathians). Studii şi Cercetări-Meteor (I): 587–599 (in Romanian)

Raţ T (1973) Durata perioadei de topire a stratului de zăpadă pe teritoriul României (Snowmelt duration in Romania). Cul. Lucr. Meteor./1970: 193–196 (in Romanian)

Schröter D, Cramer W, Leemans R, Prentice C, Araújo MB, Arnell NW, Bondeau A, Bugmann H, Carter TR, Garcia CA, de la Vega-Leinert AC, Erhard M, Ewert F, Glendining M, House JI, Kankaanpää S, Klein RJT, Lavorel S, Lindner M, Metzger MJ, Meyer J, Mitchell TD, Reginster I, Rounsevell M, Sabaté S, Sitch S, Smith B, Smith J, Smith P, Sykes MT, Thonicke K, Thuiller W, Tuck G, Zaehle S, Zierl B (2005) Ecosystem service supply and vulnerability to global change in Europe. Science. doi:10.1126/science.1115233

Şerbu C (1968) Circulaţia aerului la Muntele Semenic (Airflow circulation in Semenic Mountains). Hidrotehnica 13(8):434–436 (in Romanian)

Sonesson M, Messerli B (eds) (2002) The Abisko Agenda: research for mountain area development, Ambio Special Report 11. Royal Swedish Academy of Sciences, Stockholm

Şorodoc C (1960) Variaţia temperaturii aerului cu altitudinea în regiunea Munţilor Bucegi (Vertical variation of air temperature with height in the Bucegi Mountains). MHGA. 1

Spinoni J, Szalai S, Lakatos M, Szentimrey T, Bihari Z, Mihic D, Antofie T, Vogt J, Auer I, Hiebl J, Milkovic J, Stepanek P, Tolasz R, Zahradníček P, Nagy A, Nemeth A, Kovacs T, Kilar P, Limanowka D, Pyrc R, Cheval S, Gyorgy D, Dumitrescu A, Matei M, Birsan MV, Dacic M, Petrovic P, Krzic A, Antolovic I, Nejedlik P, Statsny P, Kajaba P, Bochnicek O, Galo D, Mikulova K, Nabyvanets Y, Skrynyk O, Krakovska S, Gnatiuk N (2014) Climate of the Carpathian Region in 1961–2010: climatologies and trends of ten variables. Int J Climatol. doi:10.1002/joc.4059

Stăncălie G, Catană S, Poiana I, Flueraru C, Căciunescu V (2004) Evaluation and monitoring of snow cover water resources in Carpathian basins using geographic information and satellite data. Rom J Meteorol 6(1–2):59–69

Stăncescu I, Damian D, Şerbu C (1986) Unele particularităţi în repartiţia precipitaţiilor atmosferice în zona Carpaţilor Româneşti şi a regiunilor limitrofe datorate proceselor foehnale (Particularities of precipitation distribution in the Romanian Carpathians and their limitrophe regions induced by foehn processes). Studii şi Cercetări-Meteor: 215–223 (in Romanian)

Stoenescu ŞM (1951) Clima Bucegilor (The climate of the Bucegi Mountains). Edit. Tehnică, Bucureşti, p 219 (in Romanian)

Stoenescu ŞM (1956) Condiţiile meteorologice favorabile formării avalanşelor (Favorable meteorological conditions for snow avalanche failures). MHGA I(1):51–54, in Romanian

Stoenescu ŞM (1959) Date noi referitoare la clima RPR (Updates in the climate of Romania). Probl Geogr VI:39–45 (in Romanian)

Stoenescu ŞM (1964) Climatic changes over the Romania's territory. RRGGG-Géogr 8:57–60

Stoenescu ŞM (1967) Resurse secundare de energie ale naturii (Alternative natural energy resources). SUBB-GG 2:175–197 (in Romanian)

Stoenescu ŞM, Dumitrescu E (1965) Frecvenţa şi variabilitatea valorilor medii zilnice ale temperaturii aerului în Depresiunea Bârsei (Frequency and variability of daily mean temperature in the Bârsa Depression). Analelele Universităţii din Bucureşti-GG, XIV (2) (in Romanian)

Ţâştea D (1959) Câteva date asupra stratului de zăpadă stabil (Stable snowpack). MHGA IV (4):18–20 (in Romanian)

Teodoreanu E (1972) Quelques considerations sur la fréquance de la hauteur maxime de precipitations en 24 heures les Carpates Méridionales. In: Proceedings of the Conf. Meteor. Carpat./1970, IMH: 492–496

Teodoreanu E (1973) Analize armonice în studiul precipitaţiilor în culoarul Rucăr-Bran (Armonic analyses applied in the study of precipitation regime in the Rucăr-Bran Corridor). SCGGG-Geogr XX(1):89–94 (in Romanian)

Teodoreanu E (1981) Le potentiel balnéo-touristique de monts de la Roumanie, avec référence spécial aux moyennes montagnes. Rev Roum Géogr Géol Géogr XXV(1):35–42

Teodoreanu E (1997) Bioclimatul din Carpaţii Româneşti şi efectele sale asupra organismului uman (Bioclimate of the Romanian Carpathians and its effects on the human body). Studii şi Cercetări-Geogr XLIV: 29–38 (in Romanian)

Tomozeiu R, Busuioc A, Ştefan S (2002) Changes in seasonal mean of maximum air temperature in Romania and their connection with large-circulation. Int J Climatol 22:1181–1196

Topoclimatologia României. Bibliografie selectivă adnotată (1987) (Topoclimatology of Romania. Adnoted selective bibliography). Institutul de Geografie. IMH, Bucureşti. p 179 (in Romanian)

Topor N (1951) Prima zăpadă (Snow onset). Ştiinţă şi tehnică pt tineret III(29):10–11 (in Romanian)

Topor N (1963) Ani ploioşi şi ani secetoşi în R.P.R. (Wet and dry years in Romania). CSA. Bucureşti: p 304 (in Romanian)

Trenberth KE, Jones PD, Ambenje P, Bojariu R, Easterling D, Klein Tank A, Parker D, Rahimzadeh F, Renwick JA, Rusticucci M, Soden B, Zhai P (2007) Observations: surface and atmospheric climate change. In: Solomon S, Qin D, Manning M, Chen Z, Marquis M, Averyt KB, Tignor M, Miller HL (eds) Climate change 2007: the physical science basis. Contribution of Working Group I to the Fourth Assessment Report of the Intergovernmental Panel on Climate Change. Cambridge University Press, Cambridge/New York

UN (United Nations) (1992) Managing fragile ecosystems: Mountain sustainable development. In: United Nations. Agenda 21

UNESCO (United Nations Educational, Scientific and Cultural Organization) (1973) Expert Panel on Project 6: impact of human activities on mountain ecosystems, Salzburg, 29, January-4 February, MAB Report No. 8. Paris

Voiculescu M (1995) Variabilitatea regimului nivometric în arealul stației Bâlea Lac – Munții Făgăraș (Variability of snoe regime in the Bâlea Lac area – the Făgăraș Mountains). Stud Cercet-Geogr XLII:113–120 (in Romanian)

Voiculescu M (2002) Incidența actuală a încălzirii climatului asupra etajului subalpin din Masivul Făgăraș (Current incidence of climate warming in the subalpine belt of the Făgăraș Massif). In: Modificările Globale ale Mediului, Lucrările celei de-a IV-a Conferințe Naționale, Comitetul Național pentru Modificările Globale, Edit. ASE: 87–96 (in Romanian)

Voiculescu M (2004a) About the morphometrical characteristics of avalanche tracks on Bâlea-Capra Area (Făgăraș Massif), International Workshop on Alpine Geomorphology and Mountain Hazards. Anal Univ de Vest din Timișoara-Geogr XIV: 31–43

Voiculescu M (2004b) Types of avalanches and their morphogenetical impact in Făgăraș Masiff – Southern Carpathians (Romania), Geomorphologia Slovaca, Císlo 1, rocník 4, Bratislava: 72–81

Voiculescu M (2009) Snow avalanche hazards in the Făgăraș massif (Southern Carpathians): Romanian Carpathians-Management and perspectives. Nat Hazards 51(3):459–475

Voiculescu M (2010) L'utilisation de la méthode dendrochronologique pour la reconstitution de la grande avalanche de neige du fevrier 1969 de Monts Bucegi – Carpates Méridionales, Roumanie, in (Surdeanu V, Stoffel M, Pop O coord.) Dendrogeomorphologie et dendroclimatologie-méthodes de reconstitution des milieux géomorphologiques et climatiques des régions montagneuses, Presa Universitară Clujeană, pp 125–149

Voiculescu M (2014) Patterns of the dynamics of human-triggered snow avalanches at the Fagaras massive (Southern Carpathians), Romanian Carpathians. Area 46:328–336. doi:10.1111/area.12111

Voiculescu M, Onaca A (2014) Spatio-temporal reconstruction of snow avalanche activity using dendrogeomorphological approach in Bucegi Mountains Romanian Carpathians. Cold Reg Sci Technol 104–105:63–75

Voiculescu M, Onaca A, Milian N, Ardelean F, Török M, Stăncescu M (2007) Analysis of snow avalanche from March 07, 2007 within the Călțun-Negoiu area, in the Făgăraș Massif (Southern Carpathians). Anal Univ din Oradea-Geogr XX(1):22–33

Voiculescu M, Onaca A, Chiroiu P (2012) L'analyse de la dynamique forestiere et de l'impact mecanique des avalanches de neige sur les arbres en utilisant la methode dendrochronologique. Étude de la vallée glaciaire Bâlea – Massif Făgăraș (Carpates Meridionales, Roumanie), Panorama de la dendrocronologie en France, Collection EDYTEM, Numéro 12, anné 2012

Zăvoianu I, Alexandrescu M, Anghel C, Mustățea A (1995) Corelații între precipitații și altitudine în Carpații Meridionali (Correlations between precipitation and height in the Southern Carpathains). Studii și Cercetări-Geogr XLII:13–19, in Romanian

Zăvoianu I, Chendeș V, Ciupitu D (1999) Factorii geografici care determină formarea și repartiția scurgerii lichide în Carpații Meridionali (Geographical factors influencing the formation and distribution of the runoff in the Southern Carpathians). Revista Geografică. VI. Serie nouă: 95–100 (in Romania)

# Chapter 3
# Study Area

**Abstract** Developed as the most extended and contorted sector of the European Alpine System, the Carpathians represent a complex natural environment whose present-day morphology is marked by active human intervention. Their limits and subdivisions were subjected to numerous geomorphic classifications, meant to separate heterogeneous units characterized by common lithological, structural or geomorphic traits. The longest sector of this European mountain range is in Romania, containing up to 40.9 % of the total Carpathian surface and 27.8 % of the country's territory. In response to a complex process of orogeny, outlined by polycyclic uplift-denudation sequences, the Romanian Carpathians Chain reflects in its morphology, morphometry and morphodynamic patterns, the litho-structural conditioning. The general physiographic disposition of the chain and the local morphometry, imposed by the folded and faulted igneous, metamorphic or sedimentary deposits, are conditioning the hydrological, climatic and biologic characteristics in terms of horizontal and vertical zonation, as well as in the temporal distribution of composing agents, processes and forms.

## 3.1 Geographical Location of the Romanian Carpathians Within the Carpathian Chain

The Carpathians are the largest, longest, most twisted and fragmented segment of the Alpine system, stretching between latitudes 50°N and 44°N, and longitudes 17°E and 27°E (Fig. 3.1). They form the eastern sector of the orographic chain that starts in the Pyrenees and continues with the Alps, and goes on across the Timok Valley towards the Black Sea, continued by the Stara Planina Mountains. Their total length is of 1,500 km and they cover a surface area of about 170,000 km$^2$ of Central-Eastern Europe, much larger than the European Alps (1,000 km and 140,000 km$^2$, respectively) (Carpathians Environment Outlook 2007).

The Romanian Carpathians is the longest sector of the Carpathians, extending along 910 km (54 % of the total length of the Carpathians) and covering an area of 66,303 km$^2$ (40.9 % of the total surface of the Carpathians and 27.8 % of Romania's territory).



D.M. Micu et al., *Climate of the Romanian Carpathians*,
Springer Atmospheric Sciences, DOI 10.1007/978-3-319-02886-6_3

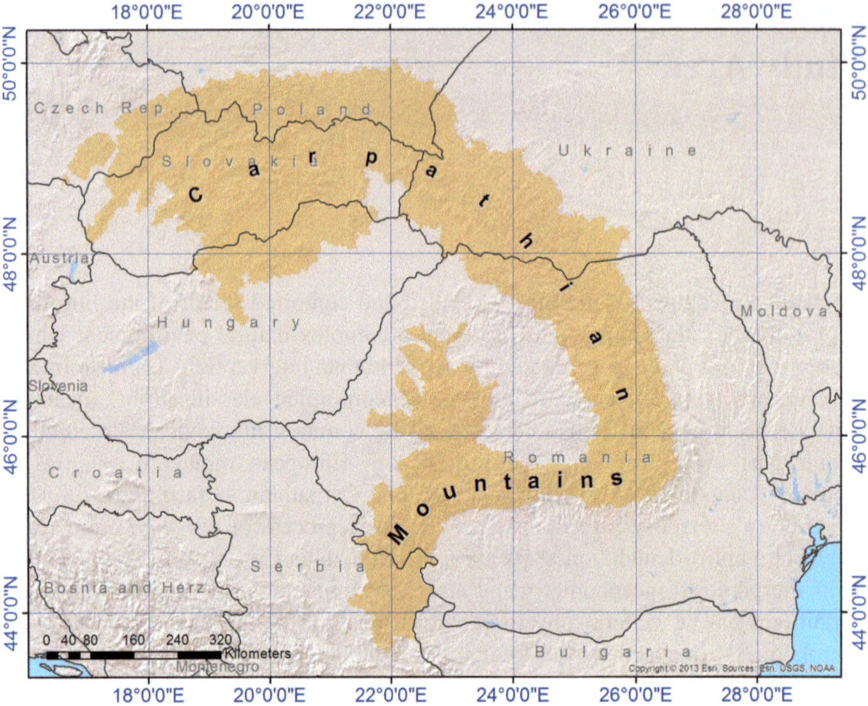


**Fig. 3.1** The location of the Romanian Carpathians within the European Carpathian chain

There are several delimitations of the Carpathian Mountains found in the specialist literature, which delineate the Romanian Carpathians using different and complex criteria: e.g. geomorphological, geological, ecological, population aspects. However, they reveal significant differentiations regarding the delimitation of the Carpathian sub-units assigned to the Romanian sector (Table 3.1). The delineations of the Romanian Carpathians existing in the national geographic literature (e.g. Posea 1972; Posea and Badea 1984; Geografia României, III, Carpații Românești și Depresiunea Transilvaniei 1987; Badea et al. 2001, 2006, 2012) are based mostly on geomorphological (morphology and morphometric traits) and geological criteria (structure, lithology).

This work uses delineations published in the geographical synthesis work Geografia României, III, Carpații Românești și Depresiunea Transilvaniei (1987). Accordingly, the Romanian Carpathians are subdivided into three main units, each of them displaying distinct physical geographical settings (Fig. 3.2, Table 3.2): the Eastern Carpathians (33,584 km$^2$), the Southern Carpathians (15,000 km$^2$), the Western Carpathians (17,714 km$^2$).

**Table 3.1** Delimitations of the European Carpathian chain (Based on the survey published by the *Carpathians Environmental Outlook* 2007 with additions)

| Carpathian divisions | Criteria | Contributions |
|---|---|---|
| The Occidental Carpathians, the Oriental Carpathians and the Transylvanian Alps | n/a | Levasseur (1886) |
| The Tatra, the Maramureş and the Transylvanian Alps | Geomorphological, human activities | George and Tricart (1954) |
| The North-Western Carpathians, the Median Carpathians and the South-Eastern Carpathians, which include the Eastern Carpathians, the Southern Carpathians and the Western Carpathians | Geomorphological, population aspects | Mihăilescu (1963) |
| The North-Western Carpathians, the North-Eastern Carpathians, the Eastern Carpathians and the Southern Carpathians | Geomorphological, geological | Székely (1968) |
| The Western Carpathians (Outer and Inner), Eastern Carpathians (Outer and Inner), Southern Carpathians and the Western Romanian Carpathians. The Carpathian area includes the surrounding hilly regions and intra-mountainous depressions | Geomorphological, geological | Kodracki (1978) |
| The North-Western Carpathians, the North-Eastern Carpathians, the Eastern Carpathians, the Southern Carpathians and the South-Western Carpathians | Geomorphological, geological, population aspects | Carpathians Environment Outlook (2007) |
| The Western Carpathians, the Eastern Carpathians and the Southern Carpathians (between the Prahova and Timok Valleys) | Geological settings and geodynamical evolution | Haass (2012) |

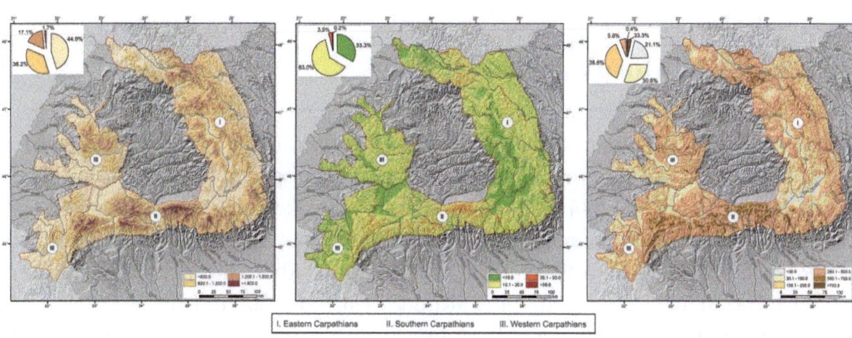


**Fig. 3.2** Main morphometric characteristics of the Romanian Carpathians: (**a**) hypsometry (m); (**b**) slope (°); (**c**) relative relief (m)

**Table 3.2** The main physical geographical settings of the Romanian Carpathians (Geografia României, III, Carpaţii Româneşti şi Depresiunea Transilvaniei 1987; Urdea 2004; Bălteanu 2012)

| Carpathian branches | Distinct elements |
|---|---|
| The Eastern Carpathians (*Rom.* Carpaţii Orientali) | Altitudes: 908 m (average), 2,303 (maximum – Pietrosu Rodnei Peak, the Rodna Mountains) |
| | Densely fragmented (25 % share of depression areas) |
| | Support extensive coniferous forest (25 %) (CLC 2006) |
| | Isolated remnants of Würm glaciation in the north, at over 2,000 m (the Rodna Mountains) |
| | Structured in three longitudinal units: crystalline in the central part, sedimentary in the outer eastern part and a Neogene volcanic in the western part |
| | Includes the longest extinct volcanic chain in Europe |
| The Southern Carpathians (*Rom.* Carpaţii Meridionali) | Altitudes: 1,071 m (average), 2,544 m (maximum – Moldoveanu Peak, the Făgăraş Mountains), 10 % share of areas higher than 2,000 m |
| | Most massive, with high relative relief values (over 80 % above 500 m) |
| | Frequent remnants of Würm and Riss glaciation at over 2,000–2,200 m |
| | Extended levelled surfaces (denudation levels) |
| | Asymmetric North-South profile (steep northern slope, gentle southern slope) |
| The Apuseni and Banat Mountains (*Rom.* Carpaţii Occidentali) | Altitudes: 652 m (average), 1,849 m (maximum – Curcubăta Mare Peak, the Vlădeasa Mountains) |
| | Densely fragmented by depressions and valleys |
| | Asymmetric West-East profile (steep eastern slope, gentle western slope) |
| | Forest area (65 %) (CLC 2006) |

## 3.2 General Morphological Settings

The Romanian Carpathians have an average elevation of 877 m (as computed from the SRTM90 m), which is well below that of the European Alps (1,350 m). Their maximum height is of 2,544 m in the Moldoveanu Peak (the Southern Carpathians), which is well below the elevation of the highest Carpathian peak – the Gerlachovsky Peak, 2,655 m (the North-Western Carpathians). About 80 % of their surface is below 1,200 m and only isolated the altitudes exceed 2,500 m (Fig. 3.2).

The mountain sector is roughly delineated from the lowlands by the 800 m elevation (Geografi a României, I, Geografie Fizică 1983). The complex morphology of the Romanian Carpathians is generally outlined by high values of fragmentation density (0.5–5.0 km/km$^2$), average relative relief of 500–600 m (exceeding 750 m, especially in the Southern Carpathians) and average slope values of 10–30° (70 % of the slopes) (Fig. 3.2).

The characteristics of the relief are an important issue when discussing climate and weather, especially through their morphometric characteristics. Altitude, slope angle, exposition (aspect) and relative height are features outlining the relief's evolution tendencies and the intensity of present-day modelling processes. Moreover, the same parameters may imprint the local character of different processes and phenomena, individualizing different morphoclimatic zones (belts). Adapting the scheme of geoecological belts defined by Kotarba (1987) in the Polish Tatras, Urdea and Sîrbovan (1995) recognized four morphoclimatic belts within the Romanian Carpathians, which are reflected in the vertical climatic and vegetation zonation: the *seminival belt* (above 2,250–2,300 m), widely corresponding to the nival belt in the Alps, the *alpine belt* (between 2,000 and 2,250 m), the *subalpine belt* (between 1,750–1,800 m and 2,000 m) and *the forest belt* (below 1,750–1,800 m). Investigating the Holocene morphogenetic altitudinal zones in the Carpathians, Kotarba and Starkel (1972), assigned the first three belts to the *cryonival morphoclimatic system* and the latter, to the *temperate forest system*. Each belt exhibit distinct prevailing geomorphic processes defining the characteristics of the actual morphodynamics in the Romanian Carpathians e.g. frost weathering, gravitational and cryogenic processes, wind erosion, and nivation are the most common processes within the subnival and alpine belts; nivation, cryogenic process, runoff and fluvial processes, creep and debris flows prevail within the subalpine belt, whereas debris flow, runoff, avalanche activity in the forest one.

The structural relief, represented by folded (in all Carpathian branches) and faulted (only in the Southern and Western Carpathians) strata may imprint a preferential development of the river network, which alongside fragmentation depth and density and overlapping regional and local circulation of the air masses may lead to local changes within important climatic parameters (temperature, precipitation and relative humidity). The petrographic relief is involved in determining local climatic features through some characteristics such as: increased heights, numerous ridges and crests, deeply-incised valleys (Southern Carpathians); endokarst and exokarst (the Apuseni Mountains); enlarged valleys and extended depressions (the extended and well-defined flysch sector of the Eastern or the less extended one of the Banat Mountains) (Fig. 3.3).

**Fig. 3.3** Structural and sculptural relief: the Siriu Mountains, Eastern Carpathians (*left*); Cindrel Mountains, Southern Carpathians (*right*) (Photo: Dana Micu)

**Fig. 3.4** Glacial and fluvial relief: the Retezat Mountains, Southern Carpathians (*left*); the Danube's Iron Gate gorges, Western Carpathians (*right*) (Photo: Dana Micu)

The climate-controlling role of the volcanic relief is very well outlined by the presence and impact of the low-altitude volcanic chain of the inner Eastern Carpathians, which through their reduced heights allow humid oceanic airflows to supply large precipitation amounts to the windward slopes compared to leeward ones. The sculptural relief (Fig. 3.4) exerts its role through the extended three level complexes, developed between 800–1,200 m, 1,300–1,600 m and 1,800–2,400 m, which are distinctive by some particular topoclimatic conditions (e.g. increased incoming solar radiation fluxes, strong airflow dynamics, enabling frequent snow transport processes in winter), which are timing the frequency and intensity of cryo-nivation processes.

Although not that widespread, the glacial relief (especially in the Southern Carpathians), through its elements (steep slopes, round-bottom long valleys and minor concavities) may favour (or just the opposite) large snow accumulations and their potential transformation into névé deposits. Despite their massive profile, the Romanian Carpathians are often transversally crossed by a dense river network.

## 3.3 The Paleogeographic Evolution and Geological Constitution

The Romanian Carpathians, part of the great Alpine-Himalayan orogeny, represent one of its most complex segment, generated by and still under the direct influence of numerous tectonic micro-plates (Pannonian, Transylvanian and Moesic). The complexity of this chain resides especially in the heterogeneous lithology marked by crystalline schists, flysch, volcanic and neo-volcanic folded and faulted structures separated by Quaternary-sediments filled intra-montane depressions. Its complex morphogenesis, outlined by the studies of Mutihac and Ionesi (1974), Mutihac (1982), Săndulescu (1984) extends over two main epochs, a pre-Carpathian one (Precambrian, Caledonian and Hercynian orogeny phases) and a Carpathian one

(Kimmerian, Old-Carpathian and Neo-Carpathian orogeny phases), which defines the actual units from a morphological, structural and lithological points of view. The last phases of evolution are outlined by the Quaternary glacial sculpting, which shapes the surfaces above 1,700–1,800 m forming a typical glacial relief, presently marked by the lack of active glaciers due to their lower altitudes. The Holocene shaping of both slopes and channels imprints the actual character of the relief, individualizing also the mountainous topoclimates.

Due to the long interval of outlining, structuring and defining its individuality, resulting from the complexity of environment and evolution processes, the Romanian Carpathians now show much structural and lithological heterogeneity. The Eastern branch features an alignment of three parallel strips formed of Cretaceous, Palaeocene and Neogene magmatites in the interior, followed by median crystalline schists (same age) and bordered towards the exterior by an extended area built on Cretaceous and Palaeogene flysch. The Southern Carpathians are characterized by N-S and E-W faulted nappe structures, consisting of crystalline core formations covered by sedimentary rocks. The western sector is the most heterogeneous one, showing an alternation of crystalline, magmatic (including Neogene magmatic rocks) and sedimentary formations, resulting in rich ferrous, non-ferrous or coal deposits.

## 3.4 Hydrology and Hydrogeology

The hydrology of the Romanian Carpathians reflects the morphogenetic, structural and petrographic complexity of this range. The large majority of Romania's rivers spring from the Carpathian Mountains and drain into the Danube. The rivers cross the Carpathians transversally, and their drainage patterns pay tribute to the position of the mountain chain in front of the prevailing humid air circulation. Gâştescu (2014) emphasized the influence of the moist (oceanic) air masses on the flow across the Romanian Carpathians. The western-facing slopes of the Apuseni Mountains (the Western Carpathians) and of the Călimani-Gurghiu-Harghita volcanic chain (the Eastern Carpathians) exhibit maximum discharge and variations of altitudinal gradients of 5–6 l/s $m^2$/100 m, while on the eastern slopes of the Eastern Carpathians, the flow is two to three times lower, due to their exposure to the prevailing continental air flows and foehn effects. The mean annual discharge at the high elevations of the Apuseni Mountains (the Western Carpathians), Retezat and Făgăraş Mountains (the Southern Carpathians), may reach 20–40 l/s $km^2$ (Geografia României, I, Geografie Fizică 1983). The Romanian Carpathians have a total water resource volume of 26.48 billion $m^3$, counting 65.3 % of the river-resources volume in Romania (Gâştescu 2014). Having a general snow-fed and pluvial regime, the rivers have a maximum discharge induced by torrential rains (during the warm season) and during spring showers which overlap the snowmelt period. Taking into account the hydropower potential, reservoirs were built on numerous mountain river courses, leading to small scale (yet locally important) microclimate changes. The distribution of ground waters (both shallow and confined aquifers) depends on differently-inclined slopes or less-to-more fissured rock formations: shallow aquifers are richer in the Intra-Carpathian depressions, while confined aquifers present very obvious lithologically-conditioned discontinuities.

## 3.5 Zonation of Vegetation and Soils

The density and floristic composition of the vegetation cover are essential to generating distinctive topoclimatic and microclimatic features in the mountainous area. Both vegetation and climate are strongly linked, so that if one is changing, the other develops variability in this causality relation. Barry (2008) stated that even "small topographic irregularities and differences of slope angle and aspect produce marked contrasts in vegetation due to the combined effects of radiation, evaporation, wind speed and snow accumulation". Climate features strongly influence the vegetation types' spatial distribution yet not exclusively. Mountain topography, edaphic conditions, latitude, etc. are also factors particularly involved in the vertical zonation of vegetation across the Carpathians. In particular, the vegetation distribution in mountain regions is so closely linked to the climatic variables that vegetation belt typology has been widely used to delineate climatic influences (e.g. Troll 1968, 1973; Lauer 1979; Monasterio 1980; Quezel and Schevok 1981; Klötzli 1984, 1991, 1994).

The Romanian Carpathians have the same characteristics common to all the other mountains belonging to the Alpine chain. Their position marks the presence of Eurasian, European, Continental, Arctic-Alpine and Alpine elements (Geografia României, I, Geografie Fizică 1983). Taking a comparative look at the bioclimatic zonation of the Alps, Tatra Mountains (the highest mountain ranges of Central Europe) and the Carpathians, several differences arise (Kotarba 1987; Ozenda 1985, 1988; Voiculescu 2002). The forest belt in the Carpathians (up to 1,600–1,800 m) largely corresponds to that of the Tatra Mountains (900–1,500 m) and to the lower and median belts of the Alps (1,000–2,200 m); the Carpathian subalpine belt (1,700–1,800 m up to 2,000 m) can be generally assigned to the Tatra (1,650–2,150 m) and to the upper mountain belt of the Alps; the alpine belt of the Romanian Carpathians (above 2,000–2,200 m) corresponds in part to the seminival belt of Tatra (2,250–2,300 m) and to the lower subnival belt of the Alps (from 2,250 up to 2,400–3,000 m).

The major types of vegetation formations in the Romanian Carpathians are distributed within four main units, among which mixed associations may appear. The nemoral belt extends between 800 and 1,200 m (sparsely 1,400 m and even 1,600 m especially along some south-facing slopes) being dominated by beech (*Fagus sylvatica*). Above it, up to 1,700–1,800 m the coniferous belt develops, formed mainly of spruce (*Picea abies*) and fir (*Abies alba*) and incidentally pine (*Pinus rubra*). The sub-alpine belt borders the upper altitudes, in the form of sparse patches, the coniferous belt (up to about 2,000 m) consists mainly of associations of dwarf spruce and pine (*Pinus mugo*), juniper (*Juniperus communis*) and Ericaceae species. Vertical zonation ends up with the alpine belt (above 2,000–2,200 m) that consists of short-grass meadow associated with dwarf shrubs.

Strongly correlated with vegetation distribution and climatic conditions, the soils follow a similar altitudinal distribution pattern. Generally, the Carpathians are characterized by soils of small-to-medium depth, poor in nutrients, with a rough texture due to intense weathering. The largest surfaces are covered with cambisols (brown eu-mesobasic, acid brown and red soils), poorly-developed soils having a very good drainage. Across the high mountain areas spodisols (podzolized brown and ferri-illuvial soils, haplic podzols and cryptopodzols) and umbrisols are developing, both intensely drained and showing a coarse texture and highly skeletal. The large depressions and floodplains are covered with luvisols (varied texture and moderate-to-good natural drainage) and hydrisols (poorly drained medium textures).

## References

Badea L, Niculescu G, Roată S, Buza M, Sandu M (2001) Unităţile de relief României, I, Carpaţii Meridionali şi Munţii Banatului (Relief units of Romania, VI, the Southern Carpathians and the Banat Mountains). Edit. Ars Docendi. Bucureşti. p 151 (in Romanian)

Badea L, Niculescu G, Sandu M, Schreiber W, Şerban M, Kadar A (2006) Unităţile de relief României, II, Munţii Apuseni şi Podişul Transilvaniei (Relief units of Romania, II, the Apuseni and Transylvania Tableland). Edit. Ars Docendi, Bucureşti. p 192 (in Romanian)

Badea L, Niculescu G, Călin D (2012) Unităţile de relief României, VI, Carpaţii Orientali – Carpaţii Maramureşului şi Bucovinei (Relief units of Romania, VI, the Eastern Carpathians – the Maramureş and Bucovina Carpathians). Edit. Ars Docendi. Bucureşti. p 106 (in Romanian)

Bălteanu D (2012) Major geomorphological units. In: Lóczy D, Stankoviansky M, Kotarba A (eds) Recent landform evolution. The Carpatho-Balkan-Dinaric Region. Springer Geography, Springer Dordrecht Heidelberg London New York, pp 250–252

Barry RG (2008) Mountain weather and climate, 3rd edn. Cambridge University Press, Cambridge, p 506

Carpathians Environmental Outlook (2007), United Nations Environment Programme/Division of Early Warning and Assessment, 232 p

CLC (2006) CORINE land cover 2006 raster data. Available at eea.europa.eu/data-and-maps/data/corine-land-cover-2006-raster-2. Accessed Sept 2013

Gâştescu P (2014) Water resources in the Romanian Carpathians-genesis, territorial distribution, management Riscuri şi Catastrofe XIII, 13(1):137–149

Geografia României, III, Carpaţii Româneşti şi Depresiunea Transilvaniei (1987) (Geography of Romania, III, the Romanian Carpathians and Transylvanian Depression). Edit. Academiei. Bucureşti (in Romanian)

Geografia României, I, Geografie Fizică (1983) (Geography of Romania, I, Physical Geography). Edit. Academiei, Bucureşti (in Romanian)

George P, Tricart J (1954) L'Europe Centrale, I, Tome Premier Géographie physique et humaine, ORBIS. Presses Universitaires de France, Paris

Haass J (2012) Geological and tectonic setting. In: Lóczy D, Stankoviansky M, Kotarba A (eds) Recent landform evolution. The Carpatho-Balkan-Dinaric Region, Springer geography. Springer, Dordrecht/Heidelberg/London/New York

Klötzli F (1984) Neuere Erkenntnisse zur Buchengrenze in Mitteleuropa, Fukarek, Akad. Nauka um jetn. bosne Herc., rad. 72, Odj. Prir. Mat. Nauka 21 [Festschr P (ed)]. Sarajevo. pp 381–395
Klötzli F (1991) Altitudinal controls on Alpine vegetation. In: Esser G, Overdiek D (eds) Longevity and stress, modern ecology: basic and applied aspects. Elsevier, Amsterdam, pp 97–110
Klötzli F (1994) Vegetation als spielball naturgegebener bauherren. Phytocoenologia 24:667–675
Kondracki J (1978) Karpaty. Wydawnictwo Szkolne i Pedagogiczne, Warsaw
Kotarba A (1987) Geoecological belts. In: Kotarba A (ed) High mountain denudational system of the polish tatra mountains. Geographical studies, Special issue (3):38–44
Kotarba A, Starkel A (1972) Holocene morfogenetic altitudinal zones in the Carpathians. Stud Geomorphol Carpatho-Balc 6:21–35
Lauer W (1979) La posicion de los páramos en la estructura del paisaje en los Andes tropicales. In: Salgado-Laboriau ML (ed) El medio ambiente páramo, actas del seminario de Mérida, Venezuela. Centro de Estudios Avanzados, Caracas, pp 29–43
Levasseur E (1886) Tracés Oro-Hydrographiques (Apennins, Karpates, Pyrénées et Plateau De Castille), in Petit Atlas de L'Europe (Moins la France) [Levasseur E, Perigot Ch (eds)]. Paris
Mihăilescu V (1963) Carpații sud-estici de pe teritoriul României. Studiu de geografie fizică cu privire specială la relief (The Southeastern Carpathians. A physical geography study focused on relief). Edit. Ştiinţifică Bucureşti. p 373
Monasterio M (1980) Los páramos andinos como region natural: caracteristicas biogeográficas generales y afinidades con otras regiones andinas. In: Monasterio M (ed) Estudios ecologicos en los páramos andinos. Ediciones de la Universidad de Los Andes, Mérida, pp 15–27
Mutihac V (1982) Unităţile geologice structurale şi distribuţia substanţelor minerale utile din România (Structural geological units and the distribution of useful minerals in Romania), Edit. Did.Pedag., Bucuresti, p 199 (in Romanian)
Mutihac V and Ionesi L (1974) Geologia Romaniei (Geology of Romania). Edit. Tehnica, Bucureşti, p 648 (in Romanian)
Ozenda P (1985) La végétation de la chaine alpine dans l'Espace montagnard Européen. Masson, Paris, p 344
Ozenda P (1988) Die Vegetation der Alpen. Gustav Fischer, Stuttgart
Posea G (1972) Regionarea carpaţilor româneşti (regionalization of the Romanian carpathians). Terra. IV(XXIV). 3 (in Romanian)
Posea G, Badea L (1984) Romania, unităţile de relief – regionarea geomorfologică (Romania, relief units – geomorphic regionalization). Map 1: 850 000, Edit. Ştiinţifică şi Enciclopedică, Bucureşti (in Romanian)
Quezel P, Schevok J (1981) Essai de mise en place de la zonation altitudinale des structures forestières. Ecologia Medit 8:389–410
Săndulescu M (1984) Geotectonica României (Geotectonics of Romania). Edit. Tehnica, Bucureşti, p 336 (in Romanian)
Székely A (1968) Európa koszorúja: a Kárpátok; Európa központi katlanja: a Kárpát-medencevilág. In: Marosi S, Sárfalvi B (eds) Európa I. kötet. Gondolat, Budapest
Troll C (1968) The cordilleras of the tropical Americas. Aspects of climatic, phytogeographical and agrarian ecology. In: Troll C (ed) Geoecology of the mountainous regions of the tropical Americas. Ferd. Dümmlers Verlag, Bonn, pp 15–56
Troll C (1973) High mountain belts between the polar caps and the equator: their definition and lower limit. Arct Alp Res 5(3):19–28
Urdea P (2004) The pleistocene glaciation of the Romanian Carpathians. In: Ehlers J, Gibbard PL (eds) Quaternary glaciations – extent and chronology. Part 1: Europe. Elsevier, Amsterdam, pp 301–308
Urdea P, Sîrbovan C (1995) Some considerations concerning morphoclimatic conditions of the Romanian Carpathians. Acta Climatol Univ Szeged 28–29:23–40
Voiculescu M (2002) Fenomene geografice de risc în Masivul Făgăraş (Geographical risk phenomena in the Făgăraş Massif) Edit. Brumar. Timişoara. p 231 (in Romanian)

# Chapter 4
# Data

**Abstract** This chapter gives information about the meteorological and ancillary datasets used in this work, including the description of the meteorological variables, their temporal coverage, the spatial distribution of the weather stations with homogenized datasets and the gridded data sources accessed (e.g. geographical, meteorological). The main steps in the development of the mountain meteorological network since the late 1920s to date and its general characteristics (e.g. number of stations, density, spatial representativity) are also summarized in a section of the chapter.

## 4.1 Datasets

The regional climatic patterns across the Romanian Carpathians were investigated based on daily time series of five key meteorological variables (Table 4.1). The original daily climatic data resulted from in situ measurements at weather stations with appropriate spatial and temporal homogeneity, which are included in the National Meteorological Administration (NMA) network. The reliability of climatic data was ensured by quality control, completion of missing data where necessary and homogenization of daily time series using the Multiple Analysis of Series for Homogenization software (MASH). The MASH homogenization procedure is further described in Chap. 5 (Sect. 5.1).

A total number of 35 weather stations, regularly distributed within the entire area of the Romanian Carpathians (15 located above 800 m and 20 below this elevation), were used to analyze the regional climatic patterns and interpret the present-day climate variability signals (Fig. 4.1). The selected sites have less than 10 % missing values over a 50-year period of observations (1961–2010) and they reflect different altitude zones and physiographic characteristics of the study region. The mountain weather stations with shorter time series were not considered in the study.



D.M. Micu et al., *Climate of the Romanian Carpathians*,
Springer Atmospheric Sciences, DOI 10.1007/978-3-319-02886-6_4

**Table 4.1** Meteorological variables used in this work

| Variables | Acronym | Description | Temporal resolution | Units | Source |
|---|---|---|---|---|---|
| Solar radiation | SID | Direct irradiance at surface | Daily | $Wm^{-2}$ | EUMETSAT's Satellite Application Facility on Climate Monitoring (CM SAF) |
| | SIS | Surface incoming shortwave radiation | Daily | $Wm^{-2}$ | |
| Air temperature | Tavg | 2 m mean daily air temperature | Daily | °C | CARPATCLIM |
| | Tmin | Minimum air temperature between 6 p.m. and 6 a.m. | Daily | °C | CARPATCLIM and in situ measurements (for absolute records) |
| | Tmax | Maximum air temperature between 6 a.m. and 6 p.m | Daily | °C | CARPATCLIM and in situ observations (for absolute records) |
| Precipitation | P | Accumulated precipitation amount from 6 a. m. and 6 p.m | Daily | mm | CARPATCLIM and in situ measurements (for absolute records) |
| | SF | Snowfalls, deduced from visual observations in the meteorological platform, between 7.00 p.m in the current day and 7.00 a.m of the next day | Daily | Days | In situ observations |
| | SS | Snow showers, depicted by snowfalls usually heavy, which can produce measurable snowpack, being characterized by a sudden beginning and ending, as well as by a highly variable intensity over short time-spans | Monthly | Days | In situ observations |

(continued)

**Table 4.1** (continued)

| Variables | Acronym | Description | Temporal resolution | Units | Source |
|---|---|---|---|---|---|
| Wind | WD | 10 m wind at eight cardinal directions | Daily | % | In situ measurements |
| | WS | 10 m wind speed | Daily | m/s | CARPATCLIM and in situ measurements (for absolute records) |
| | n/a | Strong winds deduced from readings when wind speed exceeds 16 m/s | Monthly | Days | In situ observations |
| | n/a | Atmospheric calm, when wind speed is below 1 m/s | Monthly | % | CARPATCLIM |
| Snowpack | HS | Snow depth resulting from the average reading of three fixed measuring stakes, at 6 a.m. | Daily | cm | In situ measurements |
| | SWE | The water amount resulting from the melting of a snowpack measuring at least 5 cm and covering at least 1 $m^2$, which is determined according to the formula: SWE=snow density [$g/cm^3$]*snow depth [cm]*10. | 5 days | $l/m^2$ | In situ measurements |

Large-scale circulation patterns over Europe (70°N–40°N and 20°W–40°E) were studied using charts depicting the 500 hPa geopotential height and upper air pressure, 850 hPa air temperature, front analysis, severe weather indices (e.g. Lifting Index, CAPE) and surface weather elements (e.g. 2 m air temperature, precipitation form, 10 m wind speed) retrieved from the Wetterzentrale [http://wetterzentrale.de/topkarten/] and Wetter3 websites [http://wetter3.de]. The historical Global Forecast System charts have been used to estimate the typical flow patterns and the synoptic background corresponding to certain climate extreme episodes across the Romanian Carpathians e.g. extreme cooling, warming, heavy rainfalls, gusty wind.

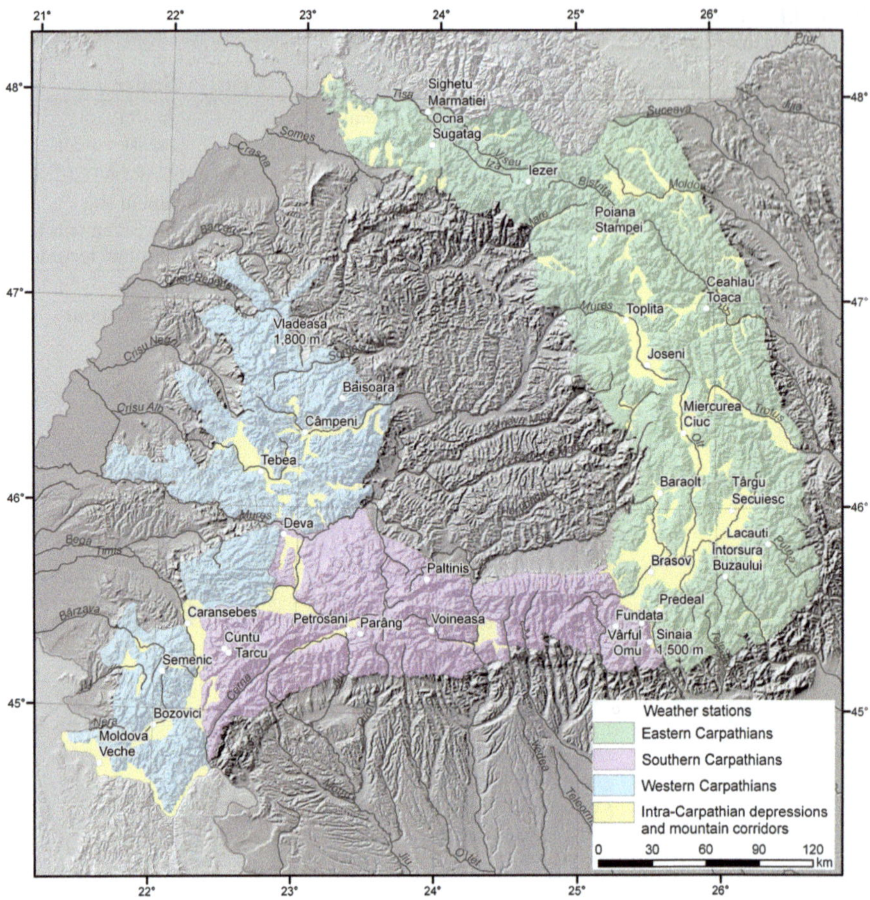


**Fig. 4.1** Spatial distribution of the weather stations used in this work

Ancillary gridded datasets have been used in this work to describe the main topographic and vegetation cover features of the Romanian Carpathians:

- Elevation data above mean sea level: the Digital Terrain Model SRTM-90 [http:\\srtm.usgs.gov\], available at a medium resolution of 1 arc-s in latitude and longitude for the entire Romanian Carpathian territory; a number of topographic parameters have been extracted from the DTM (i.e. altitude, slope, aspect) and used to investigate the vertical zonation and the local and regional climatic patterns across the study region.
- Land cover data: CORINE (Coordination of Information on the Environment) Land Cover (CLC) 2006 database derived from SPOT-4 and IRS LISS III satellite images obtained from 2006, available at 100 m horizontal resolution (EEA 2011; http://sia.eionet.europa.eu/CLC2006).

## 4.2 Mountain Meteorological Network

The beginning of mountain climatology research is widely related to the establishment of the first weather stations in the Romanian mountain region. The first mountain meteorological observations in Romania date back to the 1925–1927 interval, when several weather stations were set up at above 800 m, particularly in the Southern Carpathians: Casa Omul (2,510 m), Casa Peştera (1,615 m), Susaiu (1,329 m), Retivoiu (1,150 m), Predeal (1,090 m), Sinaia Monastery (879 m), Gheorgheni (815 m) and Păltiniş-Sibiu (1,403 m). The highest altitude weather station in Romania and also, one of the highest in Europe, has been commissioned in 1927, in a nearby location of Casa Omu establishment (the Southern Carpathians) – Vf. Omu (2,504 m). Some of these stations still operate today, without major changes in their local environmental conditions (e.g. Predeal, Păltiniş, Vf. Omu).

The strained international political context and the war explain the reduced climatological activity in Romania over the 1934–1945 period. Consequently, observations at several weather stations, especially at those in the north of the country, were discontinued, particularly in the years 1944 and 1945.

Between 1951 and 1960 efforts were made to re-open the old weather stations and set up new ones: Parâng (1,548 m) and Fundata (1,348 m) in the Southern Carpathians; and Rarău (1,536 m) in the Eastern Carpathians. In 1952, a Decision of the Council of Ministers stipulated the conditions under which weather platforms could be located so as to ensure spatial representativity and relevance of measurements and meteorological observations. That Decision paved the way for the future implementation of the state-based meteorological network devised by the Climatology Department of the Meteorological Institute. The theoretical principles underlying it helped improving database quality.

From 1960, a task plan for the reorganization of the state-based meteorological network was being implemented,[1] recommending the selection of the most appropriate conditions for setting up weather stations and rain gauge stations throughout the country in representative spatial locations, and aligning them to the international programme of meteorological measurements. After 1960, the mountain meteorological network became progressively denser (Fig. 4.2), providing more homogeneous datasets, recorded under various local geographical conditions and elevations. The majority of the existing weather stations located above 800 m (52 %) began functioning after 1961, when the national network was reorganised (Fig. 4.3). There are only five mountain sites which have data records of more than

[1] The changes in the reading time of meteorological observations in Romania are listed below:

- Before 1961: the observers measured the meteorological variables three times a day (08:00, 14:00 and 20:00).
- Since 1961: The observers measured the meteorological variables four times a day (01:00, 07:00, 13:00 and 19:00 in winter and 02:00, 08:00, 14:00 and 20:00 in summer – local time).

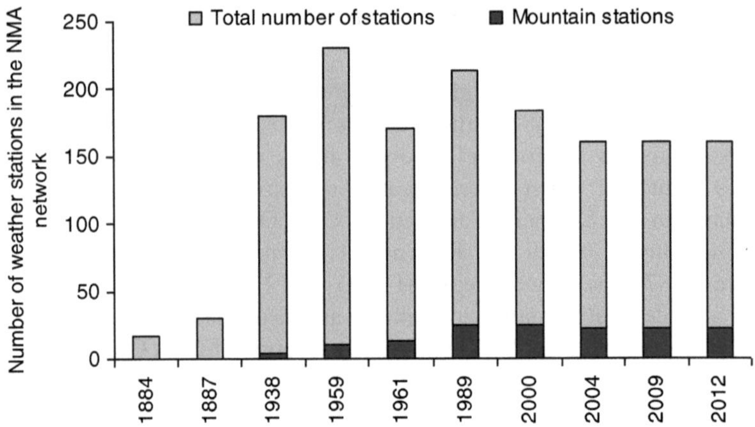


**Fig. 4.2** Development of the mountain meteorological network in Romania

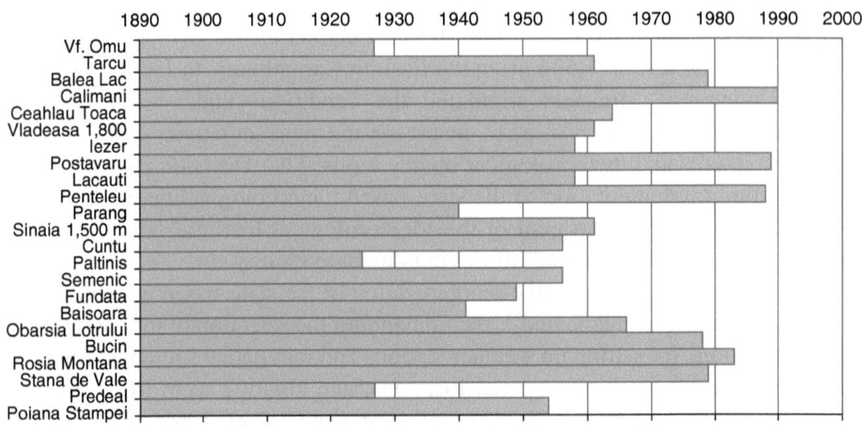


**Fig. 4.3** Commissioning years of weather stations located above 800 m

70 years: Păltiniş (89 years), Vf. Omu and Predeal (87 years), Parâng (74 years) and Băişoara (73 years).

Currently, the NMA coordinates the activity of 159 weather stations regularly distributed across Romania, among which 24 are located above 800 m altitude. Two high-elevation sites are included in the international meteorological data flow (Ceahlău-Toaca and Vf. Omu). The need to improve data quality and to provide the users with more accurate information fostered some changes in the configuration of the national network, by including automatic weather stations (as from

1995). In 2012, the NMA had a total number of 86 automatic stations operating in its network, of which 10 are located in the Romanian Carpathians.

The general features of the network spatial representativity in the Romanian Carpathian region play a key role in the analysis of regional climatic patterns. The meteorological/climatological practice defines the spatial representativity of in situ measurements as "the capacity of weather stations to supply data comparable between different regions, even though no WMO-issued standardised definition to this effect exists" (Raliţă 2006, p. 153).

Raliţă (2006) made an analysis of the degree of shelter using 149 weather stations operating in the NMA network (21 located above 800 m), with the aim of identifying potential data series error sources at observation time or in the future, derived from the local station surroundings. The extent to which the local horizon of the station is sheltered is assessed by taking into account all obstacles (natural or anthropic) found within a 300 m-distance corresponding to all 16 cardinal directions, as stipulated by WMO. The NMA assessment methodology in this respect (proposed by Raliţă 2006) is based on the distance between the centre of the platform and each obstacle found within the station's surroundings, as well as on obstacle length and height. These elements allow for determining the average vertical angle between obstacle heights by relation:

$$\alpha = arctg\left(\frac{h}{d}\right)$$

Where,

$\alpha$ = average angle by 16 cardinal directions;
h = obstacle height;
d = distance from meteorological platform centre to obstacle.

The results obtained for 21 mountain weather stations (Fig. 4.4) indicate that the mountain network consists dominantly of 'open' and 'moderately sheltered' stations, as follows:

- 38.1 % of the stations have 'open' platforms and they are located at above 1,600 m (e.g. Semenic, Lăcăuţi, Obârşia Lotrului); some of the high-elevation sites (Vf. Omu, Vf. Ţarcu, Călimani) are considered representative for the free troposphere conditions;
- 33.3 % of the stations have 'moderately sheltered' platforms, due to local obstacles that are not very tall and often even isolated (e.g. Predeal, Ceahlău-Toaca, Fundata, Băişoara);
- 28.6 % of the stations have 'sheltered' platforms (e.g. Penteleu, Păltiniş, Iezer).

The 35 weather stations, having less than 10 % missing values over the 1961–2010 period, are unevenly distributed across the three Romanian Carpathians branches (Table 4.2). These sites cover a wide elevation range, between 82 m (Moldova Veche, in the Western Carpathians) and 2,504 m

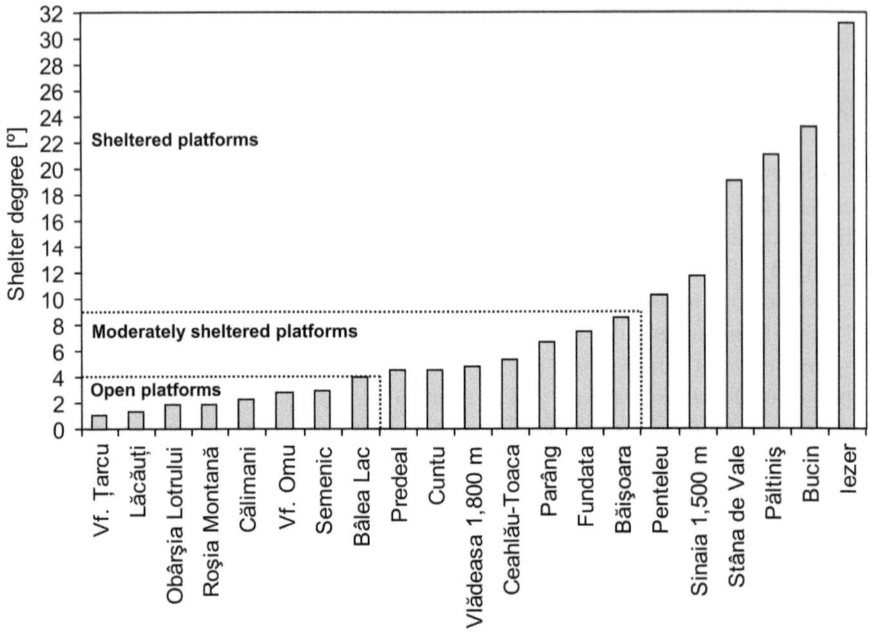


**Fig. 4.4** Platform shelter grade ($\alpha$) of weather stations in the Romanian Carpathians located above 800 m: open platforms ($\alpha < 4°$); moderately sheltered platforms ($\alpha$ is 4–9°); sheltered platforms ($\alpha$ is 9–24°)

**Table 4.2** General features of meteorological network of sites with less than 10 % missing values in the Romanian Carpathians

| Romanian Carpathian units | Surface ($km^2$) | # of NMA weather stations with less than 10 % missing values | | Overall network density (stations/50 $km^2$) |
|---|---|---|---|---|
| | | Below 800 m | Above 800 m | |
| Eastern Carpathians | 34,549 | 10 | 4 | 0.020 |
| Southern Carpathians | 14,040 | 4 | 8 | 0.042 |
| Western Carpathians | 17,714 | 6 | 3 | 0.025 |
| Total | 66,303 | 20 | 15 | 0.026 |

(Vf. Omu, in the Southern Carpathians). Their spatial distribution across the study region shows a general underrepresentation of high-elevation areas, above 1,700–1,800 m.

Table 4.3 summarizes the metadata, the data completeness and climatic variables availability for weather stations selected for conducting the current work.

**Table 4.3** List of weather stations located within the borders of the Romanian Carpathians, metadata and dataset coverage

| WMO code | Station name | Commissioning year | Geographical coordinates (decimal degrees) | Altitude (m) | Local geographical setting | Land cover/land use type | 1961–2010 dataset completeness (%) |
|---|---|---|---|---|---|---|---|
| **Eastern Carpathians** | | | | | | | |
| 656555 | Ceahlău-Toaca | 1964 | 46.97776°N, 25.95151°E | 1,897 | Mountain peak | Alpine grassland | 92 |
| 737439 | Iezer | 1958 | 47.32492°N, 25.13604°E | 1,785 | Glacial valley | Alpine grassland | 100 |
| 551621 | Lăcăuți | 1958 | 45.82401°N, 26.37708°E | 1,776 | Mountain peak | Alpine grassland | 100 |
| 719507 | Poiana Stampei | 1954 | 47.32492°N, 25.13604°E | 923 | Intra-Carpathian depression | Alpine grassland | 100 |
| 642540 | Joseni | 1921 | 46.70608°N, 25.51417°E | 750 | Intra-Carpathian depression | Arable land | 100 |
| 541601 | Întorsura Buzăului | 1946 | 45.66855°N, 26.05830°E | 707 | Intra-Carpathian depression | Build-up area | 100 |
| 655522 | Toplița | 1953 | 46.92664°N, 25.36153°E | 687 | Intra-Carpathian depression | Build-up area | 100 |
| 622544 | Miercurea Ciuc | 1948 | 46.37158°N, 25.77417°E | 661 | Intra-Carpathian depression | Arable land | 100 |
| 600608 | Târgu Secuiesc | 1954 | 45.99324°N, 26.11687°E | 568 | Intra-Carpathian depression | Arable land | 100 |
| 542532 | Braşov | 1921 | 45.69613°N, 25.52772°E | 534 | Intra-Carpathian depression | Build-up area | 100 |
| 605537 | Baraolt | 1953 | 46.08104°N, 25.59740°E | 508 | Intra-Carpathian depression | Build-up area | 100 |
| 747356 | Ocna Şugatag | 1921 | 47.77737°N, 23.94214°E | 503 | Intra-Carpathian depression | Grassland | 100 |

(continued)

**Table 4.3** (continued)

| WMO code | Station name | Commissioning year | Geographical coordinates (decimal degrees) | Altitude (m) | Local geographical setting | Land cover/land use type | 1961–2010 dataset completeness (%) |
|---|---|---|---|---|---|---|---|
| 758355 | Sighetul Marmaţiei | 1948 | 47.93957°N, 23.90588°E | 275 | Intra-Carpathian depression | Build-up area | 100 |
| **Southern Carpathians** | | | | | | | |
| 527527 | Vf. Omu | 1927 | 45.44.614°N, 25.45826°E | 2,504 | Mountain peak | Alpine grassland | 100 |
| 515231 | Ţarcu | 1961 | 45.28117°N, 22.53434°E | 2,180 | Mountain peak | Alpine grassland | 100 |
| 523328 | Parâng | 1940 | 45.38769°N, 23.46462°E | 1,548 | Mountain slope | Alpine grassland | 100 |
| 523530 | Sinaia 1,500 m | 1961 | 45.35526°N, 25.51571°E | 1,510 | Mountain slope | Forest (coniferous) | 100 |
| 518231 | Cuntu | 1956 | 45.30081°N, 22.50305°E | 1,456 | Mountain slope | Alpine grassland | 100 |
| 539357 | Păltiniş | 1925 | 45.65751°N, 23.93400°E | 1,453 | Mountain slope | Forest (coniferous) | 100 |
| 528518 | Fundata | 1949 | 45.43191°N, 25.27327°E | 1,384 | Mountain corridor | Forest (broad-leaved)/ grasslands | 100 |
| 530535 | Predeal | 1927 | 45.50646°N, 25.58510°E | 1,090 | Mountain slope | Build-up area | 100 |
| 525323 | Petroşani | 1939 | 45.40661°N, 23.37825°E | 607 | Intra-Carpathian depression | Arable land | 100 |
| 525358 | Voineasa | 1950 | 45.41150°N, 23.96855°E | 573 | Intra-Carpathian depression | Build-up area | 100 |
| 553254 | Deva | 1885 | 45.86504°N, 22.90046°E | 240 | River valley | Build-up area | 100 |

| | | | | | | | |
|---|---|---|---|---|---|---|---|
| **Western Carpathians** | | | | | | | |
| 646247 | Vlădeasa 1,800 m | 1961 | 46.75956°N, 22.79579°E | 1,836 | Mountain peak | Alpine grassland | 100 |
| 507158 | Semenic | 1956 | 45.18173°N, 22.05712°E | 1,432 | Mountain peak | Alpine grassland | 100 |
| 634322 | Băişoara | 1941 | 46.53577°N, 23.31182°E | 1,360 | Mountain slope | Build-up area | 100 |
| 622303 | Câmpeni (Bistra) | 1963 | 46.36410°N, 23.04195°E | 591 | Intra-Carpathian depression | Grassland | 94 |
| 610244 | Ţebea | 1963 | 46.16976°N, 22.72770°E | 273 | Intra-Carpathian depression | Arable land | 94 |
| 455200 | Bozovici | 1950 | 44.91865°N, 22.00774°E | 256 | Intra-Carpathian depression | Arable land | 100 |
| 525215 | Caransebeş | 1921 | 45.41756°N, 22.22684°E | 241 | Mountain corridor | Build-up area | 100 |
| 444127 | Moldova Veche | 1962 | 44.72285°N, 21.63461°E | 82 | River valley | Arable land | 96 |

# References

EEA (2011) Downloadable data about Europe's environment, European Environment Agency, Copenhagen, http://www.eea.europa.eu/data-and-maps/data

Raliță I (2006) Criterii de reprezentativitate a platformelor stațiilor meteorologice pentru evaluarea schimbărilor climatice (Representativity criteria of the meteorological platforms of weather stations for assessing climate change). National Meteorological Administration, București, p 295 (in Romanian)

# Chapter 5
# Methods

**Abstract** This chapter is organized in four sections, which refer to the homogenization algorithm applied to the meteorological data, the statistical methods, the spatialisation methods used to derive the climatological maps and the regional climate models used to derive the future changes in the climate of the Romanian Carpathians. The homogenization was done using the Multiple Analysis of Series for Homogenization (MASH v3.03) method and software and the gridding was based on the Meteorological Interpolation based on the Surface Homogenized Data Basis (MISH v1.03) software, whose results were successfully applied within the CARPATCLIM project. The main statistical methods applied in the climatic analyzes are the Mann-Kendall trend test, the Kendal-Theil slope estimation and the Spearman rank correlation coefficient. Interpolation surfaces of climate variables have been constructed using the Regression Kriging method, which combines a multivariate regression model with kriging of the regression residuals. The climate change signals until 2050 were derived from the outputs of three Regional Climate Models (RegCM3, ALADIN-Climate, and PROMES) at 25 km spatial resolution, under A1B IPCC scenario.

## 5.1 Homogenization

One of the major benefits of this work is the use of homogenous datasets for all variables, and this is the merit of the CARPATCLIM project, which provided a consistent, quality controlled, harmonized and homogenous gridded data set over the entire Carpathian Mountainous Region. Within CARPATCLIM, the meteorological data from all the Carpathian countries were quality controlled, harmonized at the national borders, homogenized, and interpolated unitarily, so that a high reliability of the resulted dataset may be claimed. Besides, the daily temporal resolution and the 0.1° (≈10 km) spatial resolution are excellent arguments for climatologic applications. The homogenization used the *Multiple Analysis of Series for Homogenization* (MASH v3.03) method and software (Szentimrey 1999, 2011; Lakatos et al. 2013), mainly due to its demonstrated performance (Venema et al. 2012). In a study comparing several widely used homogenization methods, Costa and Soares 2009 found the MASH method to be one of the most

 
D.M. Micu et al., *Climate of the Romanian Carpathians*,
Springer Atmospheric Sciences, DOI 10.1007/978-3-319-02886-6_5

comprehensive procedures for homogenization. It is a relative homogenization method which makes no a priori assumption regarding the data homogeneity, and it uses an exhaustive searching scheme to detect the most probable break and shift points in the data series from each weather station. Within MASH, data completion and quality control are performed automatically. The distribution of the examined meteorological element is taken into account for using a multiplicative (e.g. for precipitation), or additive (e.g. for temperature) model, while corrections are applied to the inhomogeneous series until no break is found. The homogenization of daily data uses the parameterisation results obtained from monthly data homogenization (Szentimrey and Bihari 2007; Szentimrey 2008). The quality of the homogenized data series is evaluated by the joint comparative mathematical examination of the original and the homogenized series systems (Costa and Soares 2009). Each country homogenized the meteorological data within its national borders ($p < 0.05$), jointly with the data from 50 km-width buffers from the neighbouring countries. The gridding was based on the *Meteorological Interpolation based on the Surface Homogenized Data Basis* software (MISH v1.03; Szentimrey and Bihari 2007). More details regarding the creation of the data sets, including the harmonization, homogenization, and interpolation, as well as the quality of the results may be found at http://www.carpatclim-eu.org/pages/deliverables/.

## 5.2 Statistical Methods

### 5.2.1 Mann-Kendall Trend Test

The local significance of trends has been analyzed with the nonparametric Mann-Kendall (MK) test on a seasonal basis. The MK test (Mann 1945; Kendall 1975) is a rank-based procedure, especially suitable for non-normally distributed data, data containing outliers and non-linear trends (Salas 1993). Generally, it is a test which investigates whether the values of a given variable tend to increase or decrease with time.

The null and the alternative hypothesis of the MK test for trend in the random variable x are:

$$\begin{cases} H_0 : \Pr(x_j > x_i) = 0.5, & j > i \\ H_A : \Pr(x_j > x_i) \neq 0.5, & \text{(two-sided test)} \end{cases} \tag{5.1}$$

The MK statistic $S$ is:

$$S = \sum_{k=1}^{n-1} \sum_{j=k+1}^{n} \operatorname{sgn}(x_j - x_k) \tag{5.2}$$

where $x_j$ and $x_k$ are the data values in years $j$ and $k$, respectively, with $j > k$, $n$ is the total number of years and sgn() is the sign function:

$$\operatorname{sgn}(x_j - x_k) = \begin{cases} 1, & \text{if } x_j - x_k > 0 \\ 0, & \text{if } x_j - x_k = 0 \\ -1, & \text{if } x_j - x_k < 0 \end{cases} \tag{5.3}$$

The distribution of $S$ can be well approximated by a normal distribution for large n, with mean zero and standard deviation given by:

$$\sigma_S = \sqrt{\frac{n(n-1)(2n+5) - \sum_{i=1}^{m} t_i(i)(i-1)(2i+5)}{18}} \tag{5.4}$$

Equation 5.4 gives the standard deviation of $S$ with the correction for ties (i.e., equal values) in data, with $t_i$ denoting the number of ties of extent $i$. The standard normal variate $Z_S$ is then used for hypothesis testing.

$$Z_S = \begin{cases} \frac{S-1}{\sigma_S} & \text{if } S > 0 \\ 0 & \text{if } S = 0 \\ \frac{S+1}{\sigma_S} & \text{if } S<0 \end{cases} \tag{5.5}$$

For a two-tailed test, the null hypothesis is rejected at significance level α, if $|Z| > Z_{\alpha/2}$, where $Z_{\alpha/2}$ is the value of the standard normal distribution with an exceedance probability α/2. In this book we chose a 10 % significance level (two-tail test).

### 5.2.2 Kendall-Theil Slope Estimate

The slope estimate $b$ is conducted with the nonparametric Kendall-Theil method (also known as Theil-Sen slope estimate) which is suitable for a nearly linear trend in the variable x and is less affected by non-normal data and outliers (e.g. Helsel and Hirsch 1992). The slope is computed between all pairs $i$ of the variable $x$:

$$\beta_i = \frac{x_j - x_k}{j-k}, \text{ with } j > k\,;\ j = 2, \ldots, n\,;\ k = 1, \ldots, n-1 \tag{5.6}$$

where $i = 1\ldots N$. For $n$ values in the time series $x$ this will result in $N = n(n-1)/2$ values of $\beta$. The slope estimate $b$ is the median of $\beta_i$, $i = 1\ldots N$.

### 5.2.3 *Spearman's Rank Correlation Coefficient*

Spearman's *rho* is a nonparametric rank-based correlation coefficient used to estimate the monotonic association between two random variables. It is computed from the difference $d$ between the ranks of independently sorted variables $x$ and $y$ (Kottegoda and Rosso 1997):

$$\rho = 1 - \frac{6\sum_{i=1}^{n} d_i^2}{n(n^2 - 1)} \tag{5.7}$$

Under the null hypothesis of no correlation between $x$ and $y$, the distribution of $\rho$ can be approximated by a normal distribution with mean μρ and variance $\sigma_\rho^2$ given by

$$\begin{cases} \mu_\rho = 0 \\ \sigma_\rho^2 = 1/(n-1) \end{cases} \tag{5.8}$$

The random variables $x$ and $y$ are considered correlated at the significance level $\alpha$ if $|\rho| > Z_{\alpha/2}/\sqrt{n-1}$ for a two-tailed test.

## 5.3 Spatialization of Climatic Information

Several works focused on the climate data spatialization in Romania (e.g. Patriche 2003, 2007, 2010; Patriche et al. 2008; Cheval et al. 2011; Dumitrescu 2012). The Geographic Information System (GIS) became a valuable tool in this matter after 2000. The COST Action 719 (2001–2008) has been dedicated to the use of GIS in climatology and meteorology studies, proposing to provide an overview of the most reliable interpolation methods with applicability in climatology. Regionalization results of climatic variable like temperature, precipitation, solar radiation, sunshine duration, wind potential, snow cover, climate indices etc. from different European countries with more or less complex orography (e.g. Spain, Norway, Portugal, Slovenia, Switzerland, France) are presented in detail in the final report of the Cost Action. Numerous experiences on meteorological information spatialization come mostly from regions of high-mountain complex topography, which also have a good spatial coverage of meteorological network (e.g. Swiss Alps, French Alps). The residual interpolation approach is shown to be more appropriate than other geostatistical interpolation methods, and therefore it was applied in the current work. The linear and multivariate regression models also provided accurate results in the European countries with complex orography.

The density of the national meteorological network in the Romanian Carpathians at above 800 m altitude is insufficient to estimate accurately the spatial distribution of meteorological information through interpolation. The regional climate patterns across the region have been retrieved by using interpolation techniques, based on the meteorological records provided by more than 120 (the number of stations is different for each climatic variable) weather stations which are currently operating in the NMA network.

A Regression Kriging (RK) approach has been used in this study to spatialize the current climate patterns across the Romanian Carpathians. RK is a multivariate interpolation method, that takes into account one or more explanatory variables with a continuous spatial distribution (DEM, satellite images, etc.). This spatial interpolation method combines the multiple regression model with the kriging of the regression errors.

In a preliminary stage of this model, the statistical relationship between predictors and predictands was validated. Using the stepwise regression method, based on Akaike criteria, the best combination of predictors was chosen, only variables that were statistically significant for the spatialisation model were further used in interpolation.

In the Regression Kriging method, the modeled spatial trend by regression is the large-scale variability of the target variable and the regression residual are the local features estimated with the help of a semi-variogram model (Hengl et al. 2007):

$$\hat{Z}(s_0) = \sum_{i=1}^{p} \hat{\beta} k \cdot qk(s_0) + \sum_{i=1}^{N} \lambda_{\mathrm{i}} \cdot e(s_i)$$

where, $\hat{\beta} k$ are estimated deterministic model coefficients, $qk$. a the values of the auxiliary variables in location $s_0$, $\lambda_i$ are kriging weights determined by the spatial dependence structure of the residual and $e(s_i)$ is the residual.

The regression coefficients are estimated using a generalized least squares model:

$$\hat{\beta}_{RK} = \left(\mathbf{q}^{\mathrm{T}} \cdot \mathbf{C}^{-1} \cdot \mathbf{q}\right)^{-1} \cdot \mathbf{q}^{\mathrm{T}} \cdot \mathbf{C}^{-1} \cdot \mathbf{z}$$

where, $\hat{\beta}_{RK}$ is the vector of estimated regression coefficients, **C** is the covariance matrix of the residuals, **q** is a matrix of predictors at the sampling locations and **z** is the vector of measured values of the target variable. After the deterministic part has been estimated (regression-part), the residuals can be interpolated with kriging and added to the estimated trend.

The general framework for obtaining the spatial distribution of climate patterns in the Romanian Carpathians is illustrated in Fig. 5.1. Mean monthly, seasonal and annual values of several climatic variables of the main variables of interest (air temperature, precipitation, wind, snow parameters) have been spatialized in this work (1961–2010).

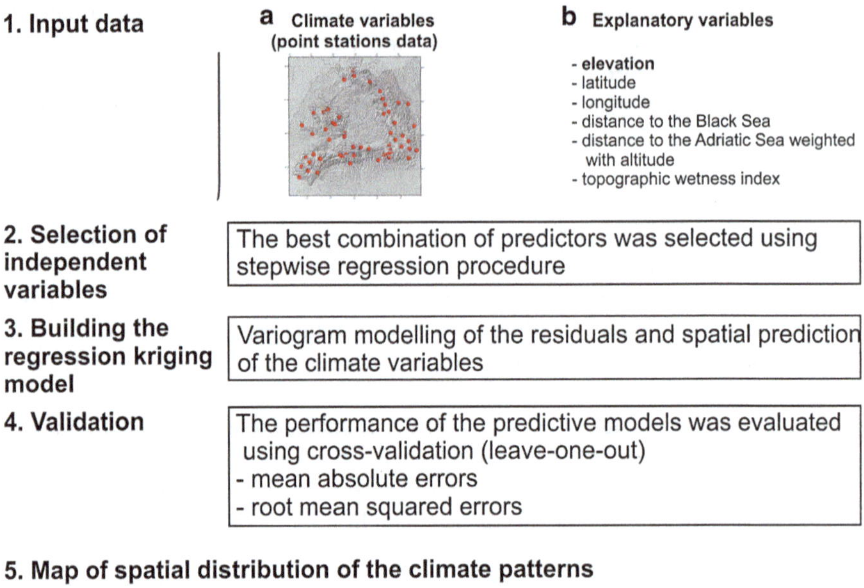


**Fig. 5.1** Framework of climatic spatialization in the Romanian Carpathians

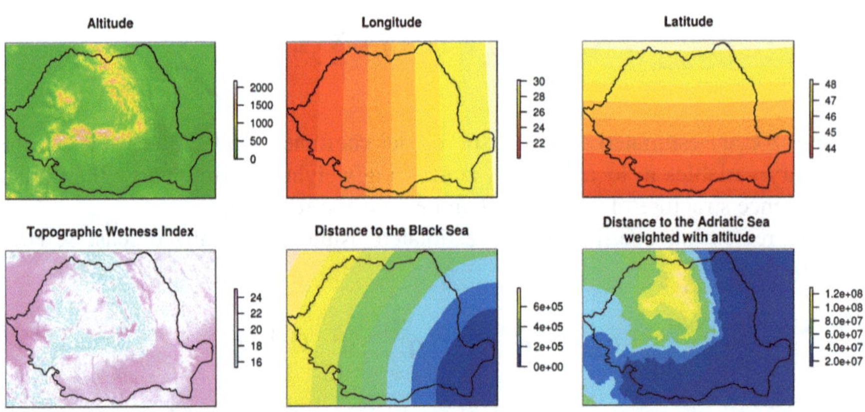


**Fig. 5.2** Spatial distribution of predictor values in Romania

Six predictors derived from a digital elevation model (DEM SRTM, 1,000 m) have been considered suitable to explain part of the spatial variability of the dependent variables (meteorological variables): altitude, location components (latitude, longitude), topographic wetness index (TWI), distance to the Black Sea and distance to the Adriatic Sea weighted with altitude (Fig. 5.2). Qin et al. (2011) emphasized the importance of TWI for quantifying the topographic control on geographical processes at hillslope or catchment scale.

The explanatory analysis of the data revealed a positively skewed distribution of the precipitation amounts, snow parameters and wind speed data. Since the spatial interpolation methods work best if the data are approximately normally distributed, all these variables were transformed using natural logarithm function. Special interest in this study has shown towards the evaluation of the relationships between the most important mountain topographic features and the main climatic attributes of these mountains (e.g. air temperature, precipitation amount, wind speed, snow cover parameters). These investigations have been carried out by using the stepwise multiple regression techniques. The findings of these analyses suggest that elevation is the most explanatory factor, as this predictor has been selected by the regression model in all cases. The model performances were slightly improved by adding the other independent variables. A summary of modelling results form some climate variables is presented in Table 5.1.

## 5.4 Regional Climate Models

While the present climate of the Romanian Carpathians is the main focus, this book also tackles the climatic projections for the next decades as such information may be essential for many stakeholders and policy makers. The investigations are based on the outputs derived from Regional Climate Models (RCMs) and available through two projects implemented within the South East Europe Transnational Cooperation Programme (SEE programme), namely CC-WaterS (http://www.ccwaters.eu), and CC-WARE (http://www.ccware.eu/). The future climate projections used in this book consider the A1B SRES IPCC scenario, and they refer to the seasonal temperature and precipitation aggregated from three RCMs (RegCM3, ALADIN-Climate, and PROMES), at 25-km spatial resolution. The analysis has been developed by comparing the average values for 1961–1990 (WMO reference period), 1991–2020 (present climate), and 2021–2050 (future climate).

The A1B SRES IPCC scenario presumes balanced energy sources within a consistent economic growth, into the context of increasing population until the mid-twenty first century, and the rapid introduction of more efficient technologies (IPCC TAR WG1 2001).

RegCM3 is the third generation of the RCM originally developed at the National Center for Atmospheric Research during the late 1980s and early 1990s. The model uses a dynamical downscaling, and it is nowadays supported by the Abdus Salam International Centre for Theoretical Physics (ICTP) in Trieste, Italy (http://regclim.coas.oregonstate.edu/statistical-downscaling/cig/index.html).

ALADIN-Climate was developed at Centre National de Recherche Meteorologique (CNRM), and it is derived from the ARPEGE-Climate as a driver for the IPCC climate scenarios over the European domain (Spiridonov et al. 2005). PROMES is a mesoscale atmospheric model developed by MOMAC (MOdelizacion para el Medio Ambiente y elo Clima) research group at the Complutense University of Madrid (UCM) and the University of Castilla-La Mancha (UCLM) (Castro et al. 1993; Gaertner et al. 2010).

**Table 5.1** Selected predictors using the stepwise regression model (marked with *), leave-one-out cross-validation statistics of the regression kriging method (mean absolute prediction error – MAE, root mean square prediction error – RMSE), and applied transformation for the target variables

| Climate variables | Explanatory variables | | | | | | Adjusted $R^2$ | MAE | RMSE | Transformation |
|---|---|---|---|---|---|---|---|---|---|---|
| | Altitude | Longitude | Latitude | TWI | Distance to the black sea | Distance to the Adriatic sea | | | | |
| Tavg annual (°C) | * | * | | * | * | | 0.970 | 0.369 | 0.494 | – |
| Tavg winter (°C) | * | * | * | * | * | * | 0.894 | 0.541 | 0.661 | – |
| Tavg spring (°C) | * | * | * | | * | * | 0.978 | 0.356 | 0.501 | – |
| Tavg summer (°C) | * | * | | | | * | 0.978 | 0.368 | 0.522 | – |
| Tavg autumn (°C) | * | * | * | * | | * | 0.959 | 0.428 | 0.558 | |
| Tmax annual (°C) | * | * | | * | * | * | 0.976 | 0.420 | 0.602 | – |
| Tmax winter (°C) | * | * | * | * | * | | 0.903 | 0.640 | 0.823 | – |
| Tmax spring (°C) | * | | * | * | * | * | 0.975 | 0.506 | 0.685 | – |
| Tmax autumn (°C) | * | * | | * | * | * | 0.976 | 0.414 | 0.603 | – |
| Tmin annual (°C) | * | * | * | * | | | 0.913 | 0.618 | 0.789 | – |
| Tmin winter (°C) | * | * | * | * | * | | 0.893 | 0.625 | 0.789 | – |
| Tmin spring (°C) | * | * | * | * | | | 0.929 | 0.603 | 0.756 | – |
| Tmin summer (°C) | * | * | * | * | | | 0.923 | 0.638 | 0.840 | – |
| Tmin autumn (°C) | * | * | * | * | * | | 0.892 | 0.682 | 0.860 | – |
| Precipitation annual (mm) | * | * | * | | * | * | 0.840 | 0.079 | 0.102 | Log |
| Precipitation winter (mm) | * | * | * | * | * | * | 0.671 | 0.113 | 0.146 | Log |
| Precipitation spring (mm) | * | | * | * | * | * | 0.837 | 0.087 | 0.111 | Log |
| Precipitation summer (mm) | * | * | * | * | * | | 0.915 | 0.078 | 0.098 | Log |
| Precipitation autumn (mm) | * | | * | * | * | * | 0.700 | 0.089 | 0.12 | Log |

| | | | | | | | | | | |
|---|---|---|---|---|---|---|---|---|---|---|
| Wind speed annual (m/s) | * | | * | | * | * | 0.555 | 0.232 | 0.318 | Log |
| Wind speed winter (m/s) | * | * | * | | * | * | 0.574 | 0.290 | 0.410 | Log |
| Wind speed spring (m/s) | * | * | * | | * | * | 0.517 | 0.219 | 0.310 | Log |
| Wind speed summer (m/s) | * | * | * | * | * | * | 0.499 | 0.213 | 0.274 | Log |
| Wind speed autumn (m/s) | * | * | * | * | * | | 0.527 | 0.252 | 0.347 | Log |
| Snow depth (cm) | * | * | | | * | * | 0.881 | 0.211 | 0.279 | Log |
| Mean snow pack duration (days) | * | * | * | | * | * | 0.858 | 0.204 | 0.265 | Log |
| Snow days | * | * | * | | * | * | 0.824 | 0.12 | 0.16 | Log |
| Snow showers days | * | * | | | * | * | 0.673 | 0.369 | 0.491 | Log |

The monthly outputs of the three models provided by the project CC-WaterS were aggregated at seasonal scale within the project CC-WARE for the whole South-Eastern Europe territory, and for this book we selected the Carpathian area.

The original 25-km spatial resolution aggregated RCM data were downscaled to 1-km in order to better emphasize the influence of the complex mountain terrain over the local climate. The centroids of the ensemble climate models were interpolated using the Regression Kriging method, based on the DEM-derived predictors described in the Sect. 5.2.

## References

Castro M, Fernandez C, Gaertner MA (1993) Description of a mesoscale atmospheric numerical model. In: Dáz JI, Lions JL (eds) Mathematics, climate and environment, vol 27, Rech Math Appl Ser. Masson, Paris, pp 230–253

Cheval S, Dumitrescu A, Petrişor A (2011) The July surface temperature lapse in the Romanian Carpathians. Carpath J Earth Environ Sci 6(1):189–198

Costa AC, Soares A (2009) Homogenization of climate data: review and new perspectives using geostatistics. Math Geosci 41:291–305. doi:10.1007/s11004-008-9203-3

Dumitrescu A (2012) Spaţializarea parametrilor meteorologici şi climatici prin tehnici SIG (Spatialization of meteorological and climatic parameters using GIS techniques). PhD thesis. University of Bucharest (in Romanian)

Gaertner MA, Dominguez M, Garvert M (2010) A modelling case-study of soil moisture–atmosphere coupling. Q J R Meteorol Soc 136(S1):483–495. doi:10.1002/qj.541

Helsel DR, Hirsch RM (1992) Statistical methods in water resources. Elsevier, Amsterdam

Hengl T, Heuvelink GB, Rossiter IDG (2007) About regression-kriging: from equations to case studies. Comput Geosci 33(10):1301–1315. doi:10.1016/j.cageo.2007.05.001

IPCC TAR WG1 (2001) Climate change 2001: the scientific basis, Contribution of Working Group I to the third assessment report of the Intergovernmental Panel on Climate Change (Houghton JT, Ding Y, Griggs DJ, Noguer M, van der Linden PJ, Dai X, Maskell K, Johnson CA, eds). Cambridge University Press, Cambridge

Kendall MG (1975) Rank correlation methods. Charles Griffin, London

Kottegoda N, Rosso R (1997) Statistics, probability and reliability for civil and environmental engineers. McGraw-Hill, New York

Lakatos M, Szentimrey T, Bihari Z, Szalai S (2013) Creation of a homogenized climate database for the Carpathian region by applying the MASH procedure and the preliminary analysis of the data. Időjárás 117(1):143–158

Mann HB (1945) Nonparametric tests against trend. Econometrica 13:245–259

Patriche C (2003) Quantifying air temperature using regression and incoming net radiation. Analele stiintifice ale Universităţii "Al. I. Cuza" din Iaşi: Geografie, 60

Patriche C (2007) About the influence of space scale on the spatialization of meteo-climatic variables. Geogr Tech 1:68–76

Patriche C (2010) Aspects concerning the outliers problem in the context of digital climatic mapping, Analele Universităţii "Ştefan cel Mare" Suceava Sectiunea. Geografie 9:5–18

Patriche C, Sfîca L, Rosca B (2008) About the problem of digital precipitations mapping using (Geo) Statistical methods in GIS. Geogr Tech 1:82–91

Qin C-Z, Zhu A-X, Pei T, Li B-L, Scholten T, Behrens T, Zhou C-H (2011) An approach to computing topographic wetness index based on maximum downslope gradient. Presion Agric 12(1):32–43

Salas JD (1993) Analysis and modeling of hydrologic time series. In: Maidment DR (ed) Handbook of hydrology. McGraw-Hill, New York

Spiridonov V, Somot S, Déqué M (2005) ALADIN-climate: from the origins to present date. ALADIN Newsletter n. 29 (Nov. 2005)

Szentimrey T (1999) Multiple Analysis of Series for Homogenization (MASH). Proceedings of the 2nd seminar for homogenization of surface climatological data. Budapest, Hungary. WMO, WCDMP-No. 41: 27–46

Szentimrey T (2008) Development of MASH homogenization procedure for daily data. Proceedings of the fifth seminar for homogenization and quality control in climatological databases, Budapest, Hungary, 2006, WCDMP-No. 71: 123–130

Szentimrey T (2011) Manual of homogenization software MASHv3.03, Hungarian Meteorological Service, p 64

Szentimrey T, Bihari Z (2007) Mathematical background of the spatial interpolation methods and the software MISH (Meteorological Interpolation based on Surface Homogenized Data Basis), Proceedings from the Conference on Spatial Interpolation in Climatology and Meteorology, Budapest, Hungary, 2004, COST Action 719, COST Office: 17–27

Venema VKC, Mestre O, Aguilar E, Auer I, Guijarro JA, Domonkos P, Vertacnik G, Szentimrey T, Stepanek P, Zahradnicek P, Viarre J, Muller-Westermeier G, Lakatos M, Williams CN, Menne M, Lindau R, Rasol D, Rustemeier E, Kolokythas K, Marinova T, Andresen L, Acquaotta F, Fratianni S, Cheval S, Klancar M, Brunetti M, Gruber C, Prohom Duran M, Likso T, Esteban P, Brandsma T (2012) Benchmarking homogenization algorithms for monthly data. Clim Past 8:89–115. doi:10.5194/cp-8-89-2012

# Chapter 6
# Geographical and Synoptic Controls on the Climate

**Abstract** This chapter outlines the importance of three major factors and the effects of their joint action in defining local and regional climatic features of the Romanian Carpathians: latitude and longitude, topography and the regional atmospheric circulation. The chapter summarizes the role of latitude and longitude in shaping the distribution of the direct and global solar radiation across the Romanian Carpathians and highlights the main influences of regional atmospheric circulation in defining local weather aspects and the regional climatic patterns in this mountain range. The complexities introduced by the main topographic characteristics of the underlying mountain terrain (e.g. altitude, slope aspect and angle, landforms) on climate state variables, are addressed here based on the key findings of several previously published works, focusing on local climate contrasts and climate zonation aspects within the Romanian Carpathians. A major focus is on the altitude effect, imposing the overall climatic zonation, as revealed by on the key findings of several previously published works.

## 6.1 Latitude

Latitude is a fundamental control on every climate. While there are other factors that affect general climate in a given area, latitude represents one of the most important factors in determining the type of climate of a given region. The latitudinal variation of the angle to the Sun implies a latitudinal variation of the solar radiation. Therefore, latitude affects temperature by influencing the seasonal range in solar intensity.

Even if the Romanian Carpathians extend only over 4° of latitude, this dependence can be noticed in the mean values of the direct radiation. However, this latitudinal dependence is noticeably altered by the orography. The latitudinal variation is more obvious during winter (when the height angles of the Sun are smaller) and less noticeable in summer.

Two solar radiation products derived from Meteosat First Generation Satellites (Meteosat 2-7) – surface incoming solar radiation (SIS) and direct irradiance (SID) – , covering the period 1983–2005, were used here to determine the



D.M. Micu et al., *Climate of the Romanian Carpathians*,
Springer Atmospheric Sciences, DOI 10.1007/978-3-319-02886-6_6

statistical influence of latitude on the spatial distribution of solar radiation across the Romanian Carpathians. The latitude influence on the seasonal variation of the surface incoming solar radiation (SIS) and the direct irradiance (SID) was found statistically significant in the Eastern and Western Carpathians, particularly in winter. Latitude was found to explain more of the SIS distribution across the two Carpathian branches, than the SID one. The statistical analyses for winter showed that latitude could explain up to 57 % of SIS variation in the Western Carpathians and up to 80 % in the Eastern Carpathians. Comparatively, the latitude influence on winter SIS distribution in regions is up to 58 % in the Western Carpathians and up to 31 % in the Eastern Carpathians. In summer, this influence is weaker, latitude explaining statistically less than 15 % of SIS distribution in the Eastern Carpathians and up to 44 % in the Western Carpathians.

## 6.2 Topography

### *6.2.1 Altitude*

Altitude is a primary controlling factor of mountain climate, as well as the most distinguishing factor of mountain climate features, imposing the vertical zonation of all climatic elements and meteorological phenomena. The altitudinal effect is outlined hereafter, in relation to the vertical zonation of some key climatic variables (e.g. solar radiation, air temperature, precipitation, wind).

There are no exhaustive studies in the specialist literature dealing with the vertical distribution of solar radiation in the Romanian Carpathians. Most studies focused on regional and local scales (e.g. Teodoreanu 1980, 1997; Andrițoiu and Ciocoiu 1971), but they are not comprehensive in terms of solar radiation characterization by its total value, temporal repartition, spectral distribution and nature (e.g. direct, diffuse, reflected).

The national radiometric network consists in a relative small number of stations (9), including only one mountain site (Poiana Brașov, 1,026 m a.s.l), which have provided records for a very short period of time (1989–1995). The solar radiation datasets at this particular site are considered not fully representative to reflect accurately the features of solar radiation regime for a mountain location or for the Romanian Carpathians in general. Using the measurements made in the national radiometric network, Oprea (2005) made a comprehensive study on solar radiation regime in Romania. This study includes some references to the features of the radiative regime in mountain regions, based on radiation records of the Poiana Brașov site and investigates the seasonal vertical variation of solar radiation relative to the radiometric sites located in lowlands. The main

**Table 6.1** Vertical gradients of the direct solar radiation in winter and summer (According to Oprea 2005, with permission)

| | Winter gradients [$Wm^{-2}$ per 100 m] | | | Summer gradients [$Wm^{-2}$ per 100 m] | | |
|---|---|---|---|---|---|---|
| Reference stations | 9 a.m. | 12 p.m. | 15 p.m. | 9 a.m. | 12 p.m. | 15 p.m. |
| Bucharest-Poiana Braşov | 12 | 11 | 15 | 6 | 6 | 6 |
| Bucharest-Fundata | 12 | 11 | 14 | 9 | 9 | 7 |
| Fundata-Poiana Braşov | 11 | 10 | 13 | 15 | 12 | 11 |

findings of the study with respect to the altitude effect on solar radiation distribution indicate:

- The dependence of the vertical variation of the direct solar radiation on the Sun height angle and weather type; this relationship appear to be stronger in winter, due to the low Sun elevation angle and the increased air masses stability (colder and with depleted water vapor content) and weaker in summer, when the elevation effect is significantly blurred by the intense thermal convection and evapotranspiration processes; Table 6.1 summarizes these effects;
- A general decrease with height of the diffuse radiation under clear sky conditions; the scattering weight displays a notable increase in mountain areas, due to the higher water vapor air content and cloud cover frequency with altitude; the diffuse flux at about 1,200 m was found to be 15–20 % higher than in the lowlands.

Global radiation shows a general increase with altitude, in response to a roughly proportional increase of the direct incoming solar beam. This altitudinal effect is particularly obvious up to 1,500 m (Clima României 2008).

The cumulative effect of altitude and solar radiation is well reflected by thermal zonation. Several studies investigated the temperature distribution with altitude in different areas of the Romanian Carpathians. Most of these studies have been conducted mainly in climatically homogeneous small mountain areas, aiming to describe near-surface temperature variations from local up to massif scales. Relevant examples are studies of Ţâştea and Neacşa (1974), in the Bâlea-Capra area (the Făgăraş Mountains, Southern Carpathians), Voiculescu (2002), in the Făgăraş Mountains, Gaceu (2005), in the Bihor and Vlădeasa mountains (the Apuseni Mountains), Bogdan (2008), in the Southern Carpathians, which briefly discussed the annual TRL patterns, based on weather stations with different length of time-series. These studies provided fairly comparable values of the temperature lapse rates: e.g. 0.5–0.7 °C/100 m (Gaceu 2005); 0.63 °C/100 m (Voiculescu 2002). Analyzing the influence of the features of underlying terrain on mountain climate and topoclimate, Niculescu (1993) indicated that under intense direct solar irradiance and summer thermal convection, the temperature lapse rate may go up to 0.8 °C/100 m. In winter, when the terrestrial radiation becomes more intense and thermal convection weakens, temperature lapse rates drop to 0.3–0.5 °C/100 m. For the first time in the specialist literature, Cheval et al. (2011) provided a full image of temperature distribution with height in July, for the whole Romanian Carpathian territory, using the land surface temperature derived from MODIS imagery. The study region still lacks detailed and systematic analysis of the air temperature distribution in relation to its controlling topographic factors, including the altitude effect.

The atmospheric relative humidity, cloudiness and precipitation amount are highly altitude-dependent, in relation to atmospheric circulation bringing air masses with increased moisture content (Clima României, 2008). The relative humidity and the degree of nebulosity are generally increasing with height at rates of 1.0–1.5 %/100 m and 0.1 tenths/100 m, respectively (Geografia României, I, Geografie Fizică, 1983). Precipitation amounts are also typically increasing with elevation. This relationship has been investigated in several works conducted mainly on local and regional scales (e.g. Apăvăloae 1972; Niculescu 1993; Voiculescu 2002; Gaceu 2005; Bogdan 2008). One of the most extensive studies on the altitudinal effects on precipitation distribution within the Romanian Carpathians was conducted by Neamu and Teodoreanu (1972), providing the first regional and local insights into the variation of precipitation gradients for this mountain range, as a whole. The study used a 60-year data coverage period (1896–1965), provided by the existing mountain weather stations and rain gauges located between 300 and 2,500 m altitude. The results indicate a considerable spatial variation in the estimated precipitation gradients across the Romanian Carpathians and also a strong differentiation of the sub-regional gradients, dependent on local mountain station density and altitudinal ranges. Accordingly, the smallest regional precipitation gradients were estimated for the Eastern Carpathians (25 mm/100 m) and the steepest in the Western Carpathians (250–300 mm/100 m). At sub-regional scale the authors found strong differentiations. Referring to the Eastern Carpathians, they estimated a proportional increase of precipitation amounts with height in the northwestern part of the Eastern Carpathians in areas up to 1,800 m (150–200 mm/100 m) and an opposite relationship in those above 1,800 m (60–80 mm/100 m). A similar vertical distribution of precipitation has been determined for the eastern half of the Southern Carpathians, suggesting a greater increase with height in areas up to 1,300 m (100–120 mm/100 m) and a less evident one in those above 1,500 m (15–20 mm/100 m). Steeper sub-regional gradients were determined for the western half of the Southern Carpathians, ranging from 120 to 150 mm/100 m in the areas up to 1,400 m to 30 mm/100 mm in those above 1,600 m.

The altitude effect on wind speed variation in the Romanian Carpathians becomes particularly evident above timberline (generally at over 1,700–1,800 m), in response to the decreasing role of the local surface roughness. Wind speeds are greater at high-elevation sites, in conditions close to the free tropospheric flow (the westerly winds). Analyzing the characteristics and trends of wind regime in Romania and the NAO influence, Vespremeanu et al. (2012) found that altitude could explain statistically about 85 % of the vertical wind speed variation, using ten wind-exposed weather stations, located at above 1,200 m in the Carpathian region. Likewise, the lowest frequency of atmospheric calm is characteristic of the high-elevation areas above 2,300 m of the Southern Carpathians (Clima României 2008).

The Romanian Carpathians are located well-below the perennial snowline, due to their medium height and middle latitude position. Voiculescu (2000) estimated the height of the theoretical perennial snowline at about 2,655 m in the Făgăraş Mountains, where the highest elevation of the Romanian Carpathian Mountains is reached (the Moldoveanu Peak, 2,544 m a.s.l.). Snow cover in the Romanian Carpathians could persist at latest until late July, but is generally rather patchy and exceptionally longer (until the next winter), only in high-elevation concave landforms (e.g. glacial cirques).

### *6.2.2 Slope Aspect*

Apart from the altitude effect, the exposure to prevailing winds is a key topographic feature, substantially involved in the generation of local climatic features (topoclimates). Its role is particularly visible when investigating the distribution of precipitation amounts on the windward and leeward slopes in relation to the westerly influence. Typical examples in this respect are offered by the Apuseni Mountains and the Eastern Carpathians, where the windward slopes are significantly wetter than the leeward ones, frequently subject to foehn effects (Gaceu 2005; Dragotă 2006; Clima României 2008). The joint effect of exposure to the dominant westerly airflows and local altitude significantly influences the height of the condensation level (CL) and consequently, the local distribution of precipitation amounts. Gaceu (2005) found that the height of the mean condensation level on the windward side of the Apuseni Mountains is particularly low, being estimated to be located at about 1,000–1,1000 m. The height of this CL was previously confirmed by Neamu and Teodoreanu (1972) and later by Croitoru et al. (2007, 2009). For comparison, Stoenescu (1951) found a higher local CL in the Bucegi Mountains, located at about 1,600–1,800 m altitude, while Dragotă (2006), highlighted the existence of a high-elevation CL on the westward side of the Rodna Mountains (at about 1,700–1,800 m). Noteworthy, is the fact that Neamu and Teodoreanu (1972) identified also two secondary CLs: one at about 1,445 m in the western part of the Southern Carpathians (in the Vâlcan Mountains) and another one, at some 1,100–1,300 m, on the northwestern flank of the Eastern Carpathians (in the Rodna Mountains).

Niculescu (1993) emphasized the importance of mountain topography in defining local climatic contrasts, asserting that windward slopes (western and north-western) have a 2–4 % higher relative humidity, 0.4–1.0 tenth more nebulosity than the leeward slopes (generally the eastern and southeastern ones) and are up to 100 m wetter compared with leeward areas of similar elevation. This work highlighted also the role of slope exposure in determining thermal differences between sunlit and lee slopes, as well as between shaded and windward ones. The findings suggest that the contrasts between windward and leeward slopes may exceed 1–2 °C, highest values being registered within the 500–1,500 m elevation range. In winter (January), these contrasts were estimated to be higher than 3 °C.

Neacşa et al. (1972) documented the relationship between slope exposure and air temperature in the Romanian Carpathians, determining differences of 1–2 °C (reaching up to 3 °C in winter) between the shaded and sunlit slopes.

Investigating the snow regime in the Romanian Carpathians, Gugiuman and Stoian (1972), showed that snow cover could persist 30 % longer on the eastern and northern slopes compared to the southern and western ones. The most favorable conditions for long-lasting snow cover were associated to the eastern slopes of the Eastern Carpathians, more frequently exposed to arctic and polar moist air advection.

The northern slopes and the deep valley corridors, colder and more shaded, exhibit good conditions for a long-lasting snowpack. In these areas, the successive snow layers may evolve into a typically denser snowpack close to the *firn*-like structure, due to compaction and metamorphism processes. Such snow deposits never reach the melting point and could be maintained from one winter to the next (e.g. in the alpine areas of the Făgăraş Mountains). The snowmelt on northern slopes is considerably delayed relative to other slopes of similar altitudes. Voiculescu and Popescu (2011) showed that the slopes of northern orientation are in general the most favorable for skiing activities.

The areas located beyond timberline (1,700–1,800 m) of the Romanian Carpathians are dominated by the influence of the atmospheric boundary layer flow. These areas are prone to frequent snow-transport processes, the snow deposition being highly dependent on the local mountain topographic features. The interaction between wind flow patterns with the local topography during snowfalls could explain the loading of the leeward slopes (generally eastern and south-eastern), caused by fresh snow accumulations or snow drifting (Voiculescu 2002). The frequency of cornice failures and slab snow avalanches on these particular slopes of the Southern Carpathians is the highest (Moțoiu 2008). At the opposite end are the southeastern and southern (sunlit) slopes, which are subject to early snowmelts under the influence of solar radiation controlling the snow surface temperature more than air temperature, entailing frequent wet loose avalanches, most common during the late winter and early-to-mid spring (Voiculescu et al. 2011; Voiculescu and Ardelean 2012).

The lee slopes subject to foehn effects (e.g. the leeside of the Apuseni Mountains, the eastern and northeastern slopes of the Banat Mountains, the northern slope of the Făgăraş Mountains, the southern and southeastern slopes of the Curvature Carpathians) are also recognized by earlier onsets of snowmelt (Bogdan 1993).

### 6.2.3 *Slope Angle*

Slope angle notably influences the distribution of the incoming solar beam, shaping the local thermal contrasts. The influence of slope angle on the distribution of direct solar radiation in the Romanian Carpathians has been previously investigated by Stan (1950). The author reported comparative solar radiation amounts accumulated over 1 year for mountain surfaces of different slope angle and aspect.

The findings of this study indicate that for mountain slopes of about 30°, the south-facing surfaces receive the highest solar flux compared to other slopes or horizontal surfaces: southern slopes (100,440 $calcm^{-2}$ min), south-eastern and southwestern slopes (89,380 $calcm^{-2}$ min), eastern slopes (72,500 $calcm^{-2}$ min) and horizontal surfaces (89,720 $calcm^{-2}$ min). Neacşa et al. (1972) asserted that the moderately steep southern slopes (15–30°) benefit in general from highest solar insolation. Later, Oprea (2005) showed that over the year, the steep north-facing slopes receive the lowest direct solar flux. The temperature contrasts driven by differences in solar radiation distribution have a key role in the development of local wind dynamics with a diurnal cycle (Whiteman 2000; Barry 2008).

There is a good correlation between the slope angle and the frequency of snow avalanche events. According to nivo-meteorological bulletins published by the National Meteorological Administration during the 2004–2010 interval, snow avalanche failures in the monitored mountain areas of the Southern Carpathians were most frequently recorded on slopes of 30–50°.

### 6.2.4 Landforms

The complexity of topographic structure of the Romanian Carpathians is also highlighted by the alternation of convex, concave and horizontal areas, reflecting the overall fragmentation and massivity degree of these mountains, which generates distinctive regional and local climatic regimes. The convex forms (peaks), tangential to the dynamic free tropospheric airstreams, are typically the coldest, wettest and windiest mountain areas, where the cold season could last 6–7 months (Micu 2007). Concave forms (e.g. glacial valleys and cirques) have a great contribution in defining distinctive topoclimatic features. Their influence usually operates rather on a local scale (from ten-to-hundreds of meters), indicating frequent channeling of colder and wetter airflows deflected from the dominant atmospheric circulation, local thermal inversions due to cold air stagnation, persistent fog and cloud cover, long-lasting air freezing and snow cover intervals (very likely to become perennial) (Bogdan and Costea 2012). The deep and narrow mountain valleys, shaded most of the day, affected by frequent cold air drainage flows and thermal inversions, favor the development of "cold pools", whose persistence is highly dependent on the synoptic weather patterns and radiative heating.

The intra-Carpathian depressions appear as vast, low-elevation areas generally below 700–800 m, formed by a mosaic of multiple horizontal surfaces, which display distinct climatic (topoclimatic) features. Bogdan and Niculescu (2004) outlined the typical climate features of mountain depressions: high frequency of negative temperatures over the year (e.g. up to 11 months in the Giurgeu-Ciuc Depression, in the Eastern Carpathians), frequent downdraughts, cold air pooling and thermal inversions (strong vertical temperature stratification up to several hundred meters), persistent air frosts (more than 150 days per year) and snow cover intervals (up to 140–150 day per year) and a high frequency of atmospheric calm (above 50 %).

## 6.3 Regional Atmospheric Circulation

The interaction between the mountain topography and the atmospheric flow is particularly complex, mountains regions playing a significant perturbation role for the large-scale processes in relation to the overall dimension and orientation of the ranges to the prevailing airflows. Beniston et al. (1997) distinguished mountain regions as "important elements of the climate system". Barry (2008) outlined that the effects of mountain topography on the atmospheric flow is obvious at different scales: e.g. the planetary scale (particularly in the case of extensive mountain ranges), the synoptic scale and the regional/local scale.

By their geographical position within the continent, the Romanian Carpathians lie in an area of interference between five major pressure centers operating over Europe e.g. Azores Anticyclone, East-European Anticyclone, Mediterranean Cyclones, Icelandic Cyclone and Scandinavian Anticyclone. Their influence is exerted more visible on synoptic and regional/local scales. Most visible effects induced by the Romanian Carpathians Chain on atmospheric flows consist of blocking of prevailing flows and perturbing the frontal structures crossing them.

The mesoscale interactions between the Carpathian topography (orography) and regional climate circulation exert a key role in defining regional climate patterns across Romania, as well as in creating contrasting climatic features within this mountain range. Many previous studies have focused on the effects of these interactions, resulting in different weather types across this mountain range and in its surrounding lowlands (e.g. Stăncescu 1972, 1983; Stăncescu and Damian 1979; Struţu and Militaru 1974; Bâzâc 1983; Bordei-Ion 1983; Moldovan 1986; Bogdan 1999, 2001). The Romanian Carpathians could explain the outstanding climatic differentiations at national scale: a cooler and wetter climate, corresponding to the western Extracarpathian region (e.g. Banato-Crişană Plain, Banat and Crişana Hills) and the Transylvanian Depression and a warmer and drier climate, corresponding to the southern and eastern Extracarpathian regions (e.g. the Romanian Plain, Moldavian Plateau, Moldavian Subcarpathians, Getic Tableland). Noteworthy, the Southern Carpathians, stretching from west to east and displaying the great elevation range and low relief fragmentation, appear to play a major role in blocking the influence of the northerly airflows of polar or arctic origins in winter, as well as of the southerly (tropical) ones. By their geographical position and the main morphological features, they divide the country into two major regions of different climatic characteristics: wetter and colder northward and drier and warmer southward. It is worth mentioning also the role of the Eastern Carpathians in blocking the action of the northeasterly and easterly air masses during winter, when the cold and stable air masses could persist over long periods on the east side of these mountains, as well as over the eastern extra-Carpathian regions. Cold air damming and thermal stable stratification under favorable atmospheric processes (prevailing anticyclonic regime) are commonly resulting in persistent low (freezing) temperatures and foggy weather conditions, compared to the sunny and warmer high-elevation areas.

The Romanian Carpathians have two curved sectors – the Carpathian Curvature sector (as part of the Eastern Carpathians) and Carpatho-Balkan Internal Curvature, which highlight their orographic barrier role in respect to the influence of the easterly, northeasterly and westerly airflows. Cordoneanu and Banciu (1991) emphasized the importance of the Eastern Carpathians and its Curvature sector, standing as a border between the prevailing oceanic and continental climate regions across Romania and inducing visible climatic differentiation between the wetter intra-Carpathian region (Transylvania) and the visibly drier eastern extra-Carpathian regions. Bogdan and Mihai (1986), asserted the role of the Carpatho-Balkan Internal Curvature, which stands as a major obstacle in the way of the eastern and north-eastern airflows, favoring the pooling of the cold continental air (usually of polar origins), for many consecutive days over the southern extra-Carpathian regions of Romania. As a result, severe negative anomalies occur, under persistent and intense inversions, frequently associated with persistent foggy weather.

The type of the prevailing air masses, generally originating from the Euro-Atlantic region, is important in determining particularities in the regional mountain climates. The westerly (oceanic) airflows traveling from the North Atlantic towards Europe are characterized by high moisture content and exert a dominant influence on the windward slopes of the Romanian Carpathians. The Apuseni Mountains, lying perpendicular to the westerly flow direction, are the first natural orographic barrier on the Romanian territory and the second one on continental scale, after the European Alps and the Jura Mountains. The influence of westerly airflows on weather types and mountain climatic patterns is particularly obvious on the western slopes of these mountains and of the Eastern Carpathians, and to a lesser extent, on the northern slopes of the Southern Carpathians. These mountain areas exhibit changeable weather and are typically wetter most of the year, with frequent precipitation falls, mostly liquid, usually associated with a high atmospheric instability during summer. The southwesterly and southerly air masses, originating from the western and central parts of the Mediterranean Basin (warm and highly humid), are responsible for causing warmer weather, high atmospheric instability resulting in frequent rain showers and shorter cold seasons. These influences are highly visible in the Banat Mountains, as well as on the southern flank of the Southern Carpathians. The winter continental airflows, usually dry and cold, originating from higher latitudes, from the Arctic Ocean, beyond 70°N latitude (cold with a lower moisture content), are particularly important for producing cold weather on the eastward slopes of the Eastern Carpathians. During the warm season, the transport of dry continental air from the steppes of Russia and Ukraine, favor de development of prolonged dry spells on the windward (eastern) slopes of the Eastern Carpathians. Continental advection coming from North Africa and Southwest Asia, typically very warm and dry, exerts a great contribution to hot and dry weather situations within the Carpathians, particularly in the Banat Mountains and on the southern flank of the Southern Carpathians and to a lesser extent, on the southern slopes of the Curvature Carpathian sector (Eastern Carpathians).

Typical weather types and mountain climate patterns, associated with the Carpathians' orographic effect, modulating the influence of the main large-scale atmospheric circulation, are summarized below:

- *westerly* (*zonal*) *circulation* (from west to east), dominant all year round, brings visibly milder weather in winter and unstable weather in summer; this flow pattern has a great influence on the temperature and precipitation regime as follows: mild winters and cool summers, more frequent rainfalls rather than sleet and snowfalls in winter, prolonged intervals of consecutive rainy days in summer resulting in significant precipitation amounts, persistent cloud cover and fog, particularly in high-elevation areas; foehn effects on the leeward slopes;
- *polar* (*meridional*) *circulation* (from north-west, north and north-east) exerts a notable influence mainly during October-March; under the influence of this atmospheric flow pattern, mountain areas are subject to cold air invasions (mostly in the Eastern Carpathians), favoring foggy and overcast weather, frequent and intense temperature contrast between windward and leeward slopes, intense temperature inversions in the intra-Carpathian depressions, cold winters and persistent snow cover.
- *tropical circulation* (south-westerly and southerly) ensures the excessive heat transfer between tropical and polar latitudes; the influence of continental tropical flow is highly relevant in explaining the occurrence of positive extremes and warm weather in the Southern Carpathians and Banat Mountains; the climate of the mountain areas subject to this circulation type is characterized by mild winters with frequent rainfall, rapid snowmelts, prolonged dry and warm spells, with rather rare rain showers in summer, mostly associated to thermal-convection; on the contrary, the tropical circulation bringing maritime airflows is significantly involved in the distribution of peak precipitation amounts over the year (typically in June and December), as well as in the occurrence of precipitation extremes, frequently associated to retrograde cyclonic activity;
- *Atmospheric blocking* situations (the least frequent) lead to fine weather, mostly clear skies and no significant precipitation amounts in summer, but overcast weather with light rains in winter.

Figures 6.1, 6.2, 6.3, 6.4 illustrates examples of atmospheric circulation patterns explaining the occurrence of some episodes of extreme cooling, warming, heavy rainfalls and gusty wind across the Romanian Carpathians, as derived from in situ mountain observations at above 800 m:

- Extreme cooling, generating the minimum absolute temperature record of −38.0 °C/February 10, 1928 at Vf. Omu (2,504 m a.s.l., the Southern Carpathians) (Fig. 6.1).
- Extreme summer warming, giving the maximum temperature records in the areas below 1,500 m of the Banat Mountains and Southern Carpathians; this advective situation resulted in the occurrence of the absolute maximum temperature record at Predeal station (1,090 m a.s.l), in the Southern Carpathians, of 33.2 °C/July 5, 2000 (Fig. 6.2).

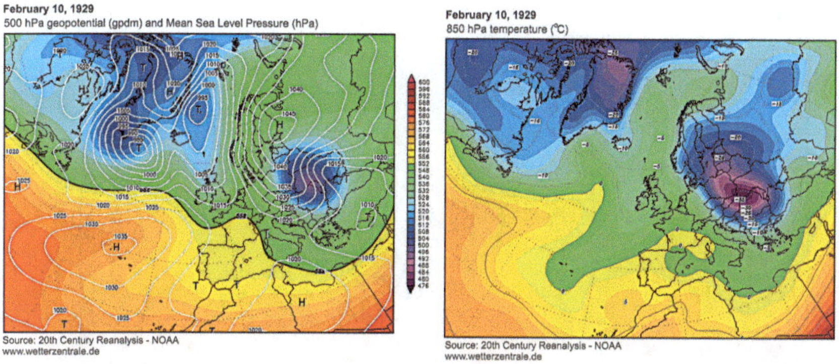


**Fig. 6.1** Continental airflows on an easterly circulation component, inducing extreme cooling in the Eastern and Central Europe

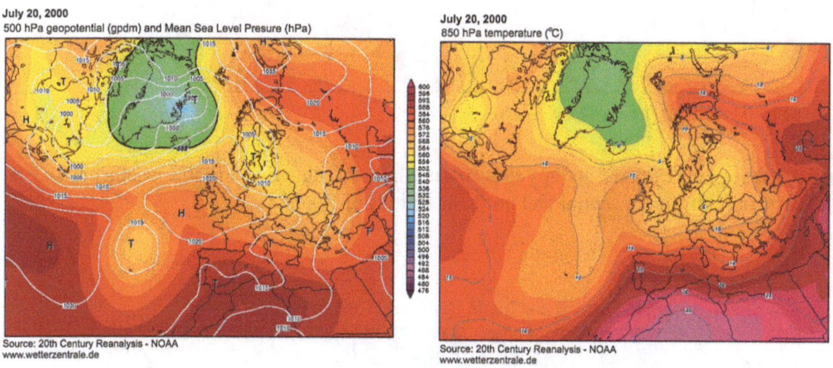


**Fig. 6.2** Tropical air advection on a south-westerly trajectory causing intense summer warming in the Banat Mountains and Southern Carpathians

- Heavy rainfalls inducing the greatest 1-day summer precipitation record at Cuntu station (1,456 m a.s.l.), in the Southern Carpathians, of 204.2 mm/July 19, 1970 (Fig. 6.3).
- Gusty winter wind episode, reaching 84 m/s (above the Beaufort 10 scale) during the night of November 27–28, 1932, in the ridge area of the Bucegi Mountains, Southern Carpathians (as reported by Stoenescu 1951) (Fig. 6.4). This extreme wind gust episode destroyed the anemometer of the Vf. Omu station.

Higher temperatures coupled with rainfall events during the snow season proved to trigger rapid snowmelts in the Carpathian region, resulting in severe floods in the lowlands. Pop and Vasenciuc (1999) investigated synoptic causes and meteorological conditions favouring the occurrence of winter floods in Romania, investigating 13 events registered over the 1989–1999 period, which affected also different areas of the Carpathian region. Accordingly, the blocking situations and westerly

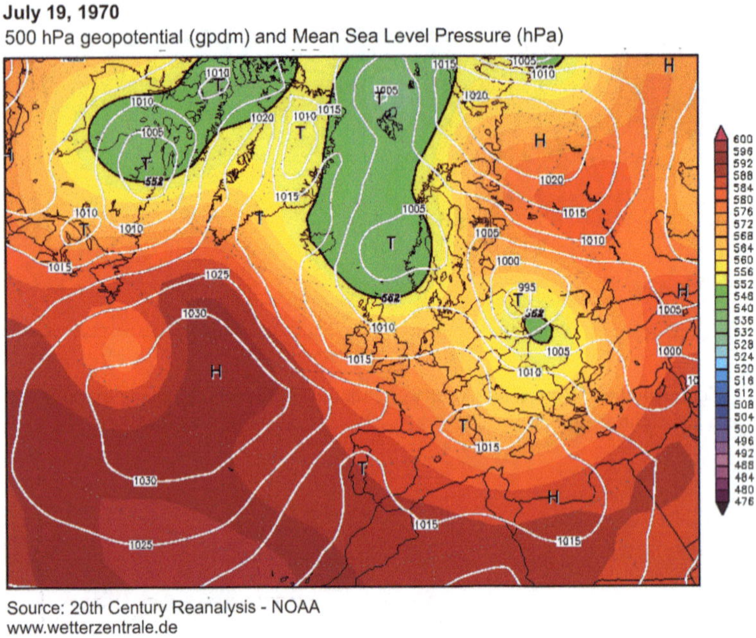


**Fig. 6.3** Retrograde cyclonic activity (towards north-west) triggering unstable weather associated with heavy rainfalls in the central and northern regions of Romania

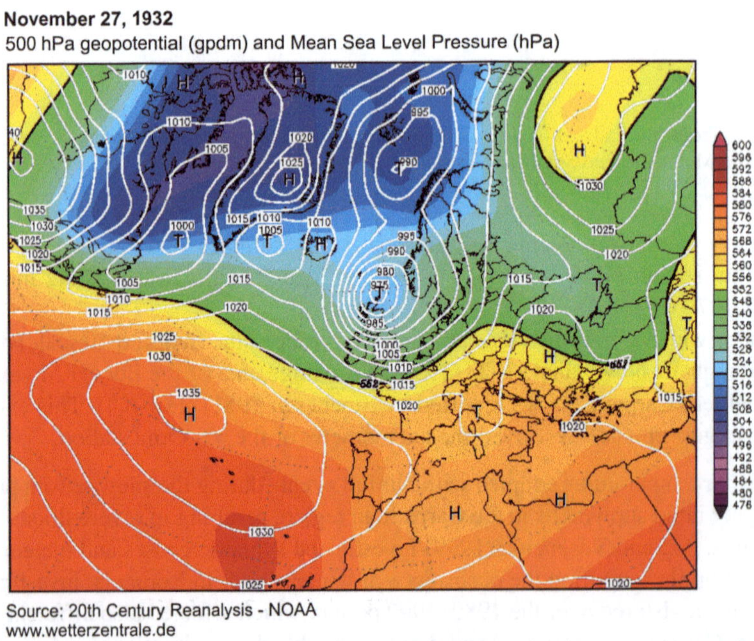


**Fig. 6.4** High geopotential gradient (about 8 hPa) along with strong temperature differences at 850 hPa level over Romania, causing gusty winds at high-elevations of the Southern Carpathians

circulations were found to count up to 40 and 46 % respectively in the generation of winter floods, while the southerly circulations only up to 12 % of the total number of cases. Several works asserted also, the contribution of warmer winter and spring weather to snow avalanche failure likelihood (e.g. Voiculescu 1996, 2002, 2004, 2009; Moţoiu et al. 2004, 2008).

By their location and arc-shape, the Romanian Carpathians play an important role in producing lee cyclogenesis on a regional scale, which usually result in rapid and notable weather changes. Bordei-Ion and Bordei-Ion (1970) demonstrated that the frequency of the Carpathian orographic cyclogenesis is particularly high during the cold-half of the year (about six cases/month in December and nine cases/ month in March). The authors showed that in winter this process is due to the influence of the Scandinavian Anticyclone, or to one of its ridges, the extension of the East-European Anticyclone towards Central Europe and the presence of a low-pressure system in the Mediterranean Basin.

Bordei-Ion (1983) gives a description of the mechanism and types of the Carpathian orographic cyclogenesis. According to him the massivity and shape of the Carpathian barrier compels the cold air mass in motion towards the north-east to skirt it and be modified (in its lower layers) by the terrestrial relief. The anticyclone forms two peri-Carpathian lobes: one over the Moldavian Plateau and the other over the Eastern Beskids, the Lower Carpathians and the Tisa Plain. In the south of Romania, or in Transylvania even, these cold advection events meet the warm air mass of a semi-stationary front (consisting in a segregation zone between the vast high-pressure area from the centre or north of the continent and the baric low in the eastern Black Sea Basin, positioned south of the Southern Carpathians, over the Getic Subcarpathians or the Getic Piedmont). Shifting southwards, the two Pericarpathian lobes decrease the temperature, triggering showers in their forefront, sleet and snowfall associated with gusty winds. The eastern lobe is the most active one, causing severe, rapidly evolving weather, while the western one is more belated and weaker in terms of temperature and dynamics". The author identified two major patterns of Carpathian orographic cyclogenesis: *complete cyclogenesis*, showing that its formation is favored by a predominantly southward trajectory of the Scandinavian Anticyclone towards the Carpathians (entailing severe cold weather in winter), and *incomplete cyclogenesis*, developed within an unevolved cyclonic structure or under an intense occlusion.

## References


Andriţoiu N, Ciocoiu I (1971) Unele determinări asupra gradientului vertical al radiaţiei solare directe în regiunile de munte (Determination of the vertical gradients of the direct solar radiation in mountain regions). Cul Lucr Meteor 1968:181–193 (in Romanian)

Apăvăloaie M (1972) Diferenţieri locale în regimul şi repartiţia precipitaţiilor atmosferice determinate de altitudine în sectorul mijlociu al văii Bistriţa (Local differentiations in the precipitation regime and its spatial distribution induced by height in the upper Bistriţa Valley). Lucr Simpoz Geogr Fiz Carpat 1972:253–261 (in Romanian)

Barry RG (2008) Mountain weather and climate, 3rd edn. Cambridge University Press, Cambridge, p 506

Bâzâc Gh (1983) Influenţa reliefului asupra principalelor caracteristici ale climei României (Influence of the relief on the main climatic characteristics of Romania). Edit. Academiei, Bucureşti, p 179 (in Romanian)

Beniston M, Diaz HF, Bradley RS (1997) Climatic change at high elevation sites: an overview. Clim Change 36:233–252

Bogdan O (1993) Foehnul carpatic (Carpathian foehn). Analele Universităţii din Oradea-Geogr, Oradea. III: 58–63 (in Romanian)

Bogdan O (1999) On the Carpathian orographic barrage in Romania. Anal Univ din Craiova-Geogr, serie nouă 2:5–12

Bogdan O (2001) Rolul de baraj orografic al Carpaţilor Româneşti (the orographic barrage role of the Romanian Carpathians). Anal Univ din Oradea-Geogr IX:5–14 (in Romanian)

Bogdan O (ed) (2008) Carpaţii Meridionali. Clima şi hazardele meteo-climatice şi impactul lor asupra turismului (The Southern Carpathians. Climate, meteo-climatic hazards and their impact on tourism). Publishing House of "Lucian Blaga" University, Sibiu, p 324 (in Romanian)

Bogdan O, Costea S (2012) Bazele teoretice ale climatologiei (theoretical background of climatology) Edit. Universităţii "Lucian Blaga". Sibiu (in Romanian)

Bogdan O, Mihai E (1986) Particularităţi climatice specifice generate de curbura internă Carpato-Balcanică ale regiunilor limitrofe româneşti (Climatic particularities induced by the inner Carpatho-Balkanic curvature in the extra-Carpathian regions). Terra. XVIII (XXXVIII) 2: 43–48 (in Romanian)

Bogdan O, Niculescu E (2004) Aspecte climatice specifice ale depresiunilor Giurgeu, Ciuc, Braşov (Specific climatic aspects of the Giurgeu, Ciuc and Braşov depressions). In: Factori şi procese pedogenetice din zona temperată, 2, serie nouă, Edit. Univ."Al. I. Cuza", pp 3–115 (in Romanian)

Bordei-Ion, E (1983) Rolul lanţului alpino-carpatic în evoluţia Ciclonilor Mediteraneeni (The role of the Alpino-Carpathic Chain related to the evolution of the Mediterranean Cyclones). Edit. Academiei. Bucureşti. p 136 (in Romanian)

Bordei-Ion E, Bordei-Ion N (1970) Bazinul Transilvaniei – centru de ciclogeneză orografică, (Transylvanian Basin – an orographic cyclogenesis centre). Hidrotehnica 15(8):437–442 (in Romanian)

Cheval S, Dumitrescu A, Petrişor A (2011) The July surface temperature lapse rate in the Romanian Carpathians. Carpathian Journal of Earth and Environmental Science 6:189–198

Clima României (2008) (The Climate of Romania). Edit. Academiei Române, Bucureşti (in Romanian)

Cordoneanu E, Banciu D (1991) O situaţie de anticiclogeneză intracarpatică (an event of intra-Carpathian anticyclonic situation). Studii şi Cercetări–Geof. 29

Croitoru AE, Moldovan F, Dragotă CS (2007) Intervale pluviometrice critice la vest de Munţii Apuseni (characteristic critical wet intervals in the western of the Apuseni Mountains). Riscuri şi Catastrofe IV:121–126 (in Romanian)

Croitoru AE, Moldovan F, Dragotă CS (2009) Considérations sur l'évolution des quantités de précipitations en 24 heures dans les Carpates Roumaines, XXIIème Colloque de L'Association Internationale de Cimatologie"Extrêmes climatiques: genèse, modélisation et impacts", Geographia Technica, Cluj-Napoca, pp 121–124

Dragotă CS (2006) Precipitaţiile excedentare din România (Excedentary precipitation in Romania), Edit. Academiei Române, Bucureşti, p 176 (in Romanian)

Gaceu O (2005) Clima şi riscurile climatice din Munţii Bihor şi Vlădeasa (The climate and climatic risks of the Bihor and Vlădeasa Mountains). Edit. Universităţii din Oradea, Oradea, p 284 (in Romanian)

Geografia României, I, Geografie Fizică (1983) (Geography of Romania, I, physical geography). Edit. Academiei. Bucureşti

Gugiuman I, Stoian R (1972) Zăpada în Carpații Românești (snowpack in the Romanian Carpathians). Lucr Simpoz Geogr Fiz Carpat 1970:283–296 (in Romanian)

NMA (National Meteorological Administration) Avalanche Bulletins (2004–2010), www.meteoromania.ro (in Romanian)

Micu D (2007) Particularități ale înghețul în aer din etajele forestier și alpin ale Carpaților Românești (Particularities of air freezing in the forest and alpine belts of the Romanian Carpathians). Revista Geografică (XIII). Serie nouă (in Romanian)

Moldovan Fl (1986) Tipuri de vreme caracteristice pentru partea de nord-vest a României (Weather types in the northwestern Romania). Studii și Cercetări-Meteor: 59–73 (in Romanian)

Moțoiu M, Milian N, Trușcă M (2004) Fenomene extreme din Munții Făgăraș – Episodul de avalanșă din 01–07 martie 2004 (Valea Bâlea) (Extreme events in the Făgăraș Mountains – the snow avalanche of March 1–7, 2004 (the Bâlea Valley). Paper presented at the annual session of the National Meteorological Administration, Bucharest (in Romanian)

Moțoiu M (2008) Avalanșele și impactul lor asupra mediului. Studii de caz din Carpații Meridionali (snow avalanches and their impact on environment. Case studies from the Southern Carpathians). Edit. Proxima. București. p 279 (in Romanian)

Neacșa O, Popa G, Popovici C, Dincă I (1972) Variația unor parametri ai temperaturii aerului în Carpații Românești (variation of some air temperature parameters in the Romanian Carpathians). Lucr Simpoz Geogr Fiz Carpat 1972:221–229 (in Romanian)

Neamu G, Teodoreanu E (1972) Repartiția precipitațiilor atmosferice în raport cu altitudinea în Carpații Românești (precipitation distribution by height in the Romanian Carpathians). Lucr Simpoz Geogr Fiz Carpat 1972:275–283 (in Romanian)

Niculescu E (1993) Influența suprafeței active montane românești asupra climatului și topoclimatului de tip carpatic (influence of the mountain underlying surface on the Carpathian climate and topoclimate). Anal Univ din Oradea-Geogr. (III): 64–68 (in Romanian)

Oprea C (2005) Climatul radiativ pe teritoriul României. O abordare geografică a radiației solare (radiative climate of Romania. A geographical approach of solar radiation). Book in digital format. ISBN: 973-0-03879. The National Meteorological Administration, București. p 250 (in Romanian)

Pop C, Vasenciuc F (1999) Cauzele meteorologice care au generat inundații în perioada de iarnă pe teritoriul României (Meteorological causes of winter flooding in Romania). Rev Geogr VI:15–20 (in Romanian)

Stan M (1950) Fluxul de insolație pe suprafețe cu orientări și înclinări diferite pentru latitudinile de 44–48° (solar radiation flux on surfaces with different aspects and slopes for areas between 44–48° N), Mem. și Studii, III(3) (in Romanian)

Stăncescu I (1972) Unele aspecte privind repartiția precipitațiilor abundente în bazinele hidrografice ale râurilor din Oltenia și Muntenia în funcție de influența condițiilor orografice (some aspects on heavy precipitation distribution in the river basins of Oltenia and Muntenia related to the orographic effect). Lucr Simpoz Geogr Fiz Carpat 1970:297–306 (in Romanian)

Stăncescu I (1983) Carpații – factori modificatori ai climei (the Carpthians as modifier factors). Edit. Științifică și Enciclopedică. București. p 139 (in Romanian)

Stăncescu I, Damian D (1979) Influența Carpaților Occidentali asupra regimului precipitațiilor atmosferice din regiunile vestică și centrală ale României (the influence of the Western Carpathians on precipitation regime in the western and central Romania). Studii și Cercetări-Meteor I:67–91 (in Romanian)

Stoenescu ȘM (1951) Clima Bucegilor (the climate of the Bucegi Mountains), Edit. Tehnică, București, p 219 (in Romanian)

Struțu M, Militaru F (1974) Carpații și rolul lor în evoluția ciclonilor cu deplasare retrogradă (the Carpathians and their role in the evolution of cyclones with retrograde tranjectories). Cul Lucr Meteor 1971:135–143 (in Romanian)

Țâștea D, Neacșa O (1974) Variația altitudinală a principalilor parametri climatici în zona Bâlea-Capra din Masivul Făgăraș (vertical zonation of the main climatic parameters in the Bâlea-Capra area, the Făgăraș Mountains). Studii de climatologie 1:105–209 (in Romanian)

Teodoreanu E (1980) Culoarul Rucăr-Bran. Studiu climatic şi topoclimatic (the Rucăr-Bran Corridor. A climatic and topoclimatic study). Edit. Academiei. Bucureşti. p 166 (in Romanian)

Teodoreanu E (1997) Bioclimatul din Carpaţii Româneşti şi efectele sale asupra organismului uman (Bioclimate of the Romanian Carpathians and its effects on human body). SCGGG-Geogr XLIV:29–38 (in Romanian)

Vespremeanu-Stroe A, Cheval S, Tatui F (2012) The wind regime of Romania – characteristics, trends and North Atlantic oscillation influences. Forum Geografic XI(2):118–126

Voiculescu M (1996) Tipologia celor mai importante fenomene geografice de risc în Masivul Făgăraş (The typology of the most important geographical risk phenomena in the Făgăraş Mountains). Analele Universităţii de Vest-Geogr VI:41–55 (in Romanian)

Voiculescu M (2000) Tipuri de limite superioare ale pădurii în Masivul Făgăraş (types of upper timberlines in the Făgăraş Massif). Analele Universităţii de Vest din Timişoara-Geogr IX:157–174 (in Romanian)

Voiculescu M (2002) Fenomene geografice de risc în Masivul Făgăraş (geographical risk phenomena in the Făgăraş Massif). Edit. Brumar. Timişoara. p 231

Voiculescu M (2004) Types of avalanches and their morphogenetical impact in Făgăraş Masiff – Southern Carpathians (Romania). Geomorphologia Slovaca. Císlo 1. rocník 4. Bratislava: 72–81

Voiculescu M (2009) Snow avalanches potential and management of Romanian Carpathians. Case study – Bâlea glacial valley (Făgăraş Massif – Southern Carpathians). In: Proceedings of the International Snow Science Workshop. Davos: 53–57

Voiculescu M, Ardelean F (2012) Snow avalanche – disturbance of high mountain environment. Case study – the Doamnei glacial valley the Făgăraş massif-Southern Carpathians, Romanian Carpathians. Carpath J Earth Environ Sci 7(1):95–108

Voiculescu M, Popescu F (2011) The management of snow avalanches in the ski areas in Southern Carpathians. Case study: Făgăraş Massif and Bucegi Mountains. In: Zhelezov G (ed) Sustainable development in mountain regions: southeastern Europe. Springer, Berlin, pp 103–120

Voiculescu M, Ardelean F, Onaca A, Török-Oance M (2011) Analysis of snow avalanche potential in Bâlea glacial area – Făgăraş Massif (Southern Carpathians – Romanian Carpathians). Z Geomorphol Stuttg 55(3):291–316

Whiteman CD (2000) Mountain meteorology. Fundamentals and applications. Oxford University Press, New York, p 355

# Chapter 7
# Regional Climatic Patterns

**Abstract** This chapter provides a description of the main climatic conditions of the Romanian Carpathians region based on the analysis of the regime of the most important climatic parameters such as air temperature, precipitation, wind and snow. The regional climatic patterns are discussed in terms of elevation effect, influence of the prevailing atmospheric circulation and seasonality, aiming at distinguishing between the climate variation particularities emerging between three main units. The analysis highlights the climatic differentiations between the areas above 800 m and those below 800 m, generally including most of the intra-Carpathian depressions. The absolute climatic records derived from in situ measurements are presented in tabular form.

## 7.1 Solar Radiation

Variations in the intensity of incident solar radiation reaching the Earth may produce changes in global and regional climate, which are both different and additional to those from anthropogenically-induced climate change. At the present period of time, solar variation impacts on regional climate seem rather significant regionally, but on a global scale are acknowledged to be considerably smaller than those due to increasing greenhouse gases.

Solar radiation is considered one of the main climatological factors, shaping the thermal regime, determining the vegetation type and its spatial distribution, determining the growing season duration, etc. By its geographical position within Europe, Romania has a moderate solar potential. Yet, the radiation regime shows visible regional differences across the country. The annual variation of radiation incidence angle in Romania induces an increase of day length between December 22 (the winter solstice) and June 22 (the summer solstice) of 6 h and 30 min in the southern part of the country and of 7 h and 40 min in the northern part (Geografia României, I, Geografie Fizică 1983).

A number of radiation studies or synthesis works carried out on a national scale, focusing on the spatial distribution of some solar radiation variables (particularly on global solar radiation and direct solar radiation), had considered the solar radiation



D.M. Micu et al., *Climate of the Romanian Carpathians*,
Springer Atmospheric Sciences, DOI 10.1007/978-3-319-02886-6_7

as an important climatological factor. It is worth mentioning the works of Stan (1950), Stoenescu (1951), Țâștea (1961), Andrițoiu (1962), Andrițoiu and Ciocoiu (1971), Diaconescu (1966), Fărcaș (1981), Bâzâc (1983), Buiuc (1984), Bădescu (1999), Oprea (2005) and Clima României (2008). Most of these studies have considered the mountain regions as of special interest. Still, there is a general lack of solar radiation and energy budget studies for the entire territory of the Romanian Carpathians.

The data processing from Meteosat First Generation Satellites (Meteosat 2–7) provided climate datasets of surface incoming solar radiation (SIS) and direct irradiance (SID) for the present study, covering the period 1983–2005. The datasets were validated by comparison with high-quality ground based measurements from 12 stations of the Baseline Surface Radiation Network (Ohmura et al. 1998), located in the main climatic regions. According to the SIS and SID products' developer (CM-SAF), high bias values were found in the Alpine and other mountainous regions, e.g. due to uncertainties in area to point comparison and errors introduced by snow coverage.[1] They may exceed 20 $W/m^2$ for daily means of SIS, and 15 $W/m^2$ for SID.

Several studies in the national specialist literature were aimed at investigating the distribution of sunshine duration, global solar radiation, and their relationship with height and cloud cover (e.g. Neacșa and Popovici 1969; Bacinschi and Neacșa 1985; Voiculescu 2002; Oprea 2005). However, the findings of most of these studies are restricted to local or national scales. Analyzing the solar energy resources of Romania, Ciocoiu (1976), cited by (Oprea 2005) have stressed the vertical variation of the annual mean global solar radiation, based on the radiation records of ten meteorological stations, located between 443 and 2,504 m altitude. The study showed a significant influence of cloud cover on solar radiation data, particularly in the mountain slope sector where the cloud cover prevails.

The *surface incoming solar radiation* (SIS), provided by the CM SAF Surface Radiation MVIRI DataSet 1.0, has been used to analyze the altitudinal and latitude effect and to characterize the seasonality of the global solar radiation across the Romanian Carpathians. SIS is defined as the radiation flux (irradiance) reaching a horizontal plane at the Earth surface in the 0.2–4 μm wavelength region, and is expressed in $W/m^2$. SIS was retrieved using the Heliosat method, which is based on the conservation of energy. The Heliosat algorithm uses reflection measurements given as normalized digital counts to determine the effective cloud albedo (Cano et al. 1986; Beyer et al. 1996; Hammer et al. 2003). Then, a clear sky model is used to calculate the solar surface irradiance based on the retrieved effective cloud albedo (Posselt et al. 2012).

The altitude effect on the surface incoming solar radiation distribution over the year is strong in most seasons (except winter) all over the Romanian Carpathians.

[1] http://www.cmsaf.eu/bvbw/generator/CMSAF/Content/Publication/SAF__CM__DWD__PUM__MVIRI__HEL__1__4,templateId=raw,property=publicationFile.pdf/SAF_CM_DWD_PUM_MVIRI_HEL_1_4.pdf

**Table 7.1** Seasonal vertical (W/m$^{-2}$/100 m) and latitude gradients (W/m$^{-2}$/deg) of surface incoming solar radiation in the Romanian Carpathians

| Romanian Carpathians branches | DJF | MAM | JJA | SON |
|---|---|---|---|---|
| *Vertical gradients* | | | | |
| Eastern Carpathians | 0.11 | −1.53 | −2.75 | −0.87 |
| Southern Carpathians | −0.13 | −0.90 | −1.96 | −0.61 |
| Western Carpathians | −0.09 | −1.89 | −3.75 | −1.33 |
| *Latitudinal gradients* | | | | |
| Eastern Carpathians | −3.91 | −2.79 | −4.71 | −6.32 |
| Western Carpathians | −2.43 | −4.31 | −12.50 | −5.91 |

The elevation explains, statistically significantly, the spatial distribution of SIS, amounting to 63–77 % in summer, 49–77 % in spring and 20–76 % in autumn. In winter, its effect is below 10 % in all the Carpathian branches. Generally, the seasonal surface incoming solar radiation decreases with height in most of the Romanian Carpathians, except for the Eastern Carpathians in winter (Table 7.1).

The latitude effect on the seasonal SIS variation appears the strongest and more important than the altitudinal one, in the Western and particularly in the Eastern Carpathians, which extend some 2–3° latitude (Table 7.1). These regions exhibit large seasonal and diurnal differences in the weather from north to south. Latitude explains statistically the seasonal SIS variation in the range of 13–80 % in the Eastern Carpathians and of 20–66 % in the Western Carpathians. Generally, in these regions the seasonal surface incoming solar radiation increases with latitude at a rate of 3–6 W/m$^{-2}$/deg in the Eastern Carpathians and 2–13 W/m$^{-2}$/deg in the Western Carpathians.

The seasonal variation of the surface incoming solar radiation across the Romanian Carpathians is shown in Fig. 7.1. Peak SIS amounts are characteristic to summer, due to the high frequency of atmospheric stability (particularly in July and August) under a prevailing anticyclonic regime, and the high solar zenith angle. In response, the sunshine duration is the greatest in this season, reaching up to 2,100–2,200 h in the areas below 800 m and 1,800 h in those above 800 m (Clima 2008). Summer SIS exceeds 220–230 W/m$^{-2}$ in the areas below 800 m of the Eastern and Southern Carpathians, and exceptional 250 W/m$^{-2}$ in some depression areas of the Banat Mountains (Western Carpathians). Comparatively, the areas above 800 m, summer SIS values are generally below 200–210 W/m$^{-2}$. In the high-elevation areas affected by intense local thermal convection processes in the afternoon, SIS drops well below 200 W/m$^{-2}$. The peak frequencies of cloud cover during wintertime (particularly in December) caused by some typical large-scale circulation patters (the intensification of cyclonic activity over the Mediterranean Sea, the southward retreat of the Azores High and the mid-latitude expansion of the Icelandic Low), as well as the low solar zenith angle explain the lowest SIS values over the year. In this season, in most of the Southern Carpathians and extended areas of the Western Carpathians, including the Banat Mountains, the southeastern Apuseni Mountains, as well as the southeastern extremities of the Eastern

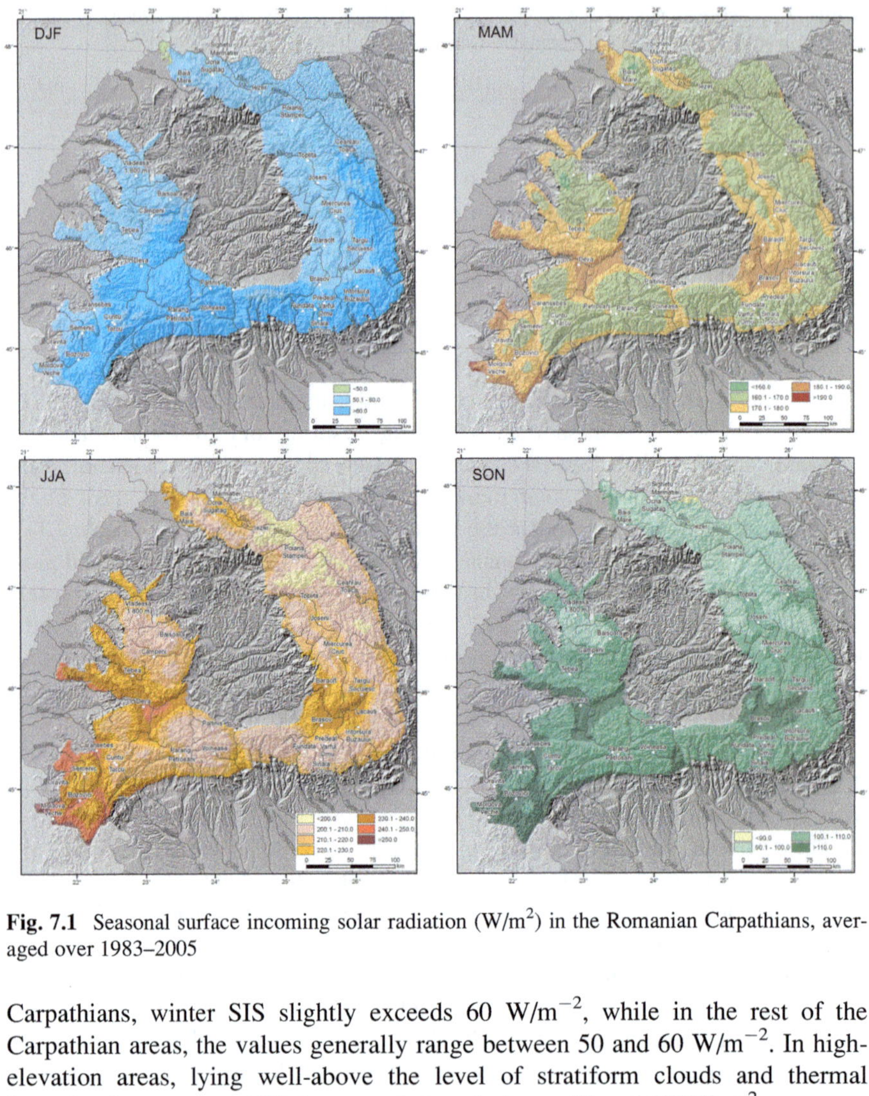


**Fig. 7.1** Seasonal surface incoming solar radiation (W/m$^2$) in the Romanian Carpathians, averaged over 1983–2005

Carpathians, winter SIS slightly exceeds 60 W/m$^{-2}$, while in the rest of the Carpathian areas, the values generally range between 50 and 60 W/m$^{-2}$. In high-elevation areas, lying well-above the level of stratiform clouds and thermal inversion layer, winter SIS shows variations between 50 and 60 W/m$^{-2}$.

The *direct irradiance* (SID) represents the radiation flux (irradiance) reaching a horizontal plane at the Earth surface in the 0.2–4 μm wavelength region directly without scattering. SID is also expressed in W/m$^2$. Ciulache (2002) asserted that the solar energy flux directly targeting the mountain active surface (unmodified), represents 21 up to 27 % of the solar constant.

Buiuc (1984) indicated that the Romanian Carpathians have the lowest direct solar radiation values at the national level (below 575 KWhm$^{-2}$), due to high cloud cover and fog frequency, emphasizing that nebulosity is one of the main atmospheric conditions that size the amount of the incoming solar radiation in the Carpathian region, with a dominant cloudy climate.

**Table 7.2** Seasonal vertical (W/m$^{-2}$/100 m) and latitude gradients (W/m$^{-2}$/deg) of surface incoming solar radiation in the Romanian Carpathians

| Romanian Carpathians branches | DJF | MAM | JJA | SON |
|---|---|---|---|---|
| *Vertical gradients* | | | | |
| Eastern Carpathians | 0.16 | −1.65 | −3.04 | −0.77 |
| Southern Carpathians | −0.28 | −1.36 | −2.81 | −0.78 |
| Western Carpathians | −0.09 | −1.96 | −4.16 | −1.06 |
| *Latitudinal gradients* | | | | |
| Eastern Carpathians | −2.30 | −2.91 | −6.10 | −4.70 |
| Western Carpathians | −1.77 | −5.86 | −15.36 | −4.65 |

The direct solar radiation in the Romanian Carpathians has a diurnal and annual variation which is typical for mid-latitudes; the diurnal variation is maintained throughout the year.

The altitude effect on the distribution of the direct irradiance over the year is strong particularly in spring and summer, when elevation explains statistically 59–69 % and 46–68 %, respectively, of the variance. This effect is maintained strongly in the Southern and Western Carpathians throughout the year. The direct irradiance decreases with height in most areas of the Romanian Carpathians, except for the Eastern Carpathians in winter (Table 7.2). In autumn and winter, the latitude influence on the SID spatial variation appears stronger than the altitude one, particularly in the Eastern and Western Carpathians. In these regions, latitude could explain statistically up to 31–58 % of winter SID variation and 59–66 % of autumn SID. Generally, in these regions the seasonal direct irradiance decreases with latitude at a rate of 2–6 W/m$^{-2}$/deg in the Eastern Carpathians and 2–15 W/m$^{-2}$/deg in the Western Carpathians.

The seasonal atmospheric circulation patterns could explain much of the seasonal SID variation over the Romanian Carpathians. In winter, the synoptic conditions drive frequent, cold, continental air masses advection and intense cyclonic activity over the Mediterranean Sea, that are favourable for producing low SID values. In this season, SID do not exceed 40 W/m$^{-2}$, ranging between 20 and 30 W/m$^{-2}$ in most of the Carpathian areas and dropping below 20 W/m$^{-2}$ only in the northern parts of the Eastern Carpathians. The late summer (August) and early autumn (September) are in general the sunniest intervals of the year across the Romanian Carpathians, due to the influence of the Azores Anticyclone, which advancing eastward favours a greater air mass stability all over Eastern Europe. This is also a result of the frequent tropical (continental) advection from North Africa or South-Eastern Asia. The prevailing atmospheric stability in summer explains the peak SID values in the Romanian Carpathians, generally ranging between 80 and 120 W/m$^{-2}$ in the areas above 800 m and exceeding 120 W/m$^{-2}$ in those below 800 m. The active thermal convection processes, forming the cumuliform cloud cover, can reduce SID amounts only locally. The seasonal variation of the direct irradiance across the Romanian Carpathians is illustrated in Fig. 7.2.

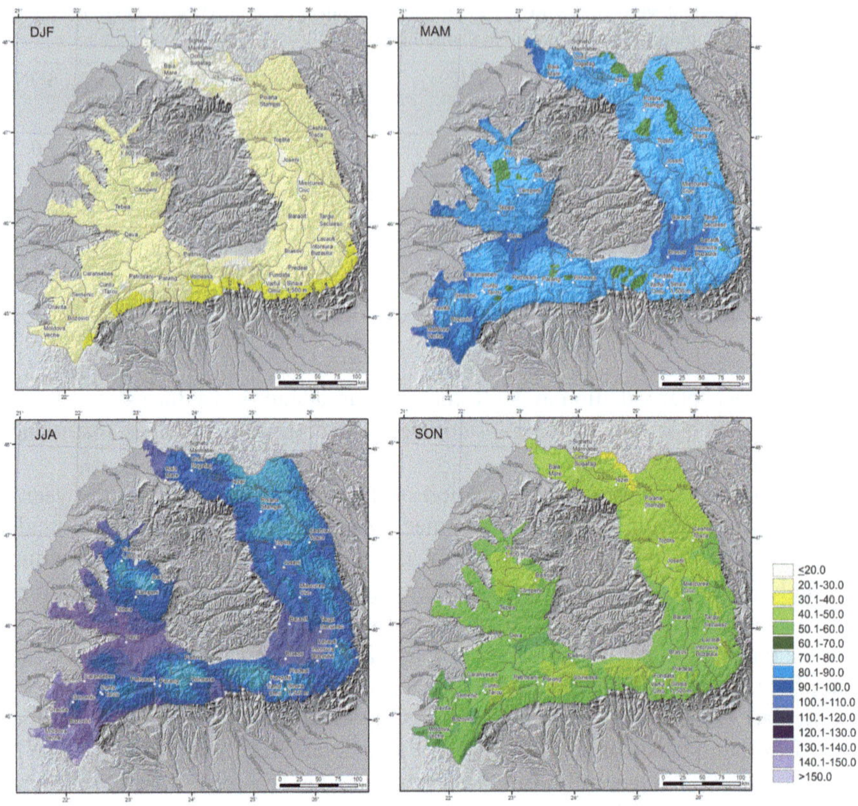


**Fig. 7.2** Seasonal direct irradiance (W/m$^2$) in the Romanian Carpathians, averaged over 1983–2005

## 7.2 Air Temperature

### *7.2.1 Average Air Temperature Lapse Rate*

Air temperature is a primary variable describing the climate of the Romanian Carpathians and an important controlling factor of many environmental processes. This variable shows large spatial variations and a great dependency upon the main features of mountain topography (slope, aspect and especially elevation). The magnitude of these variations is also a function of energy balance regimes, atmospheric moisture content, wind speed, cloudiness, radiative conditions and distance from the sea. Temperature lapse rate (TLR) is one of the most important characteristic of local and regional climate of complex terrain.

Through this study, the vertical zonation of air temperature across the Romanian Carpathians has been quantified, using a common, homogenized and long-enough period of air temperature observations, based on a multivariate geostatistic

**Table 7.3** Annual and seasonal $T_{avg}$ lapse rates (TLRs) in the Romanian Carpathians (°C/100 m)

| TLRs | Eastern Carpathians | Southern Carpathians | Western Carpathians |
|---|---|---|---|
| Annual | 0.57 | 0.62 | 0.60 |
| DJF | 0.38 | 0.47 | 0.41 |
| MAM | 0.68 | 0.74 | 0.68 |
| JJA | 0.69 | 0.72 | 0.72 |
| SON | 0.51 | 0.57 | 0.56 |

approach developed in the R programming language. The changing rate of temperature with height was determined using a 1 km resolution DEM of the Carpathian region and the gridded annual, monthly and seasonal temperature values derived from a country-wide interpolation including all the stations included in the NMA network (124). This analysis was carried out on an annual and seasonal basis for average, minimum and maximum air temperature.

Regression analyses carried out for the three Romanian Carpathian branches showed that elevation might accurately estimate the spatial distribution of the air temperature, by explaining 80 up to 95 % of its spatial variation at all time-scales considered.

An average regional lapse rate of about 0.60 °C/100 m was determined for the Romanian Carpathians territory, based on the average air temperature values ($T_{avg}$). Regionally, the variations in the TLRs are determined by the region-wide air masses characteristics over the year in relation to the elevation range (Table 7.3). The steepest lapse rate, indicating a greater decrease in temperature with height, was determined for the Southern Carpathians 0.62 °C/100 m, which cover the most extended altitude range and the smallest one, in the Eastern Carpathians −0.57 °C/100 m, which are most frequently exposed to cold continental airflows (of polar and arctic origins). An average lapse rate of 0.60 °C/100 m was found in the Western Carpathians, the wettest and warmest Carpathian region, due to the prevailing westerly and south-westerly airflows.

The annual TLRs determined in the Romanian Carpathians are fairly comparable with those reported in other studies focused on regions with complex terrains of Europe: e.g. 0.64–0.80 °C/100 m in the western Italian Alps (Cortemiglia et al. 1989); 0.57–0.61 °C/100 m in the southern French Alps (Douguédroit and De Saintignon 1970, 1981); 0.52–0.64 °C/100 m in the Jura Mountains (De Saintignon 1986); 0.54 °C/100 m in the Tyrol region (Rolland 2003).

The lapse rates estimating temperature surface conditions in the Romanian Carpathians exhibit a clear seasonal cycle, with amplitudes ranging from 0.27 °C (in Southern Carpathians) to 0.31 °C (in the Eastern and Western Carpathians). The magnitude of the TLRs change through the year is the largest in spring and summer (from 0.68 to 0.74 °C/100 m), with the greatest values in the Southern Carpathians (Fig. 7.3). In autumn and winter the decrease of air temperature with height is smaller in all the three Carpathian regions (from 0.38 to 0.57 °C/100 m). The increased frequency of cold air pooling and temperature inversions during the cold months, which may last long periods in valley-bottoms and depression areas

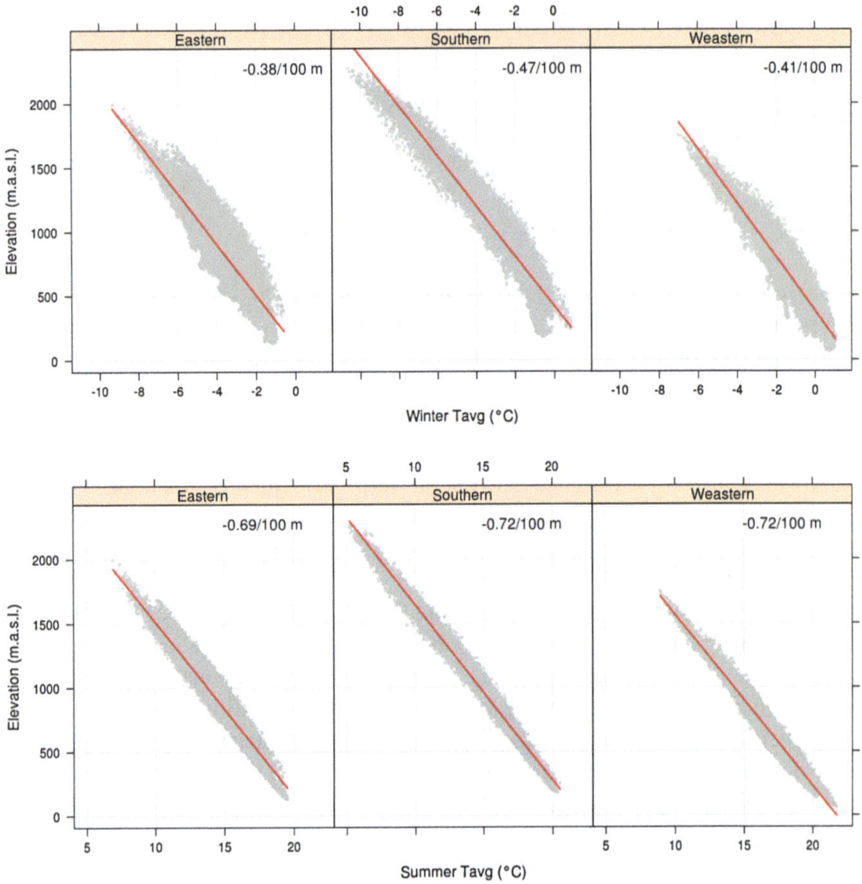


**Fig. 7.3** $T_{avg}$ lapse rates (°C/100 m) in the Romanian Carpathians during the extreme seasons

(below 800–1,000 m), alters the decrease of air temperature with height, especially during the night and can result in large and persistent local negative temperature anomalies. These phenomena explain the low magnitudes of temperature decrease with height in January, of only 0.34 °C/100 m in the Western Carpathians and 0.36–0.38 °C/100 m in the Eastern and Southern Carpathians.

### 7.2.2 *Vertical Thermal Zonation*

Hess (1965, 1971) provided the first vertical climate zonation of the Carpathians. He identified seven thermal belts, based on the relationship between the annual $T_{avg}$ and altitude and the distribution of vegetation zones. The altitudinal distribution of these belts within the Romanian Carpathians is shown in Table 7.2. According to this delineation, the 'cold' thermal belt is present only in the Southern Carpathians at above 2,100 m, due to their large altitudinal range, favoring discontinuous permafrost development, active periglacial processes and long-lasting snow cover. The Western Carpathians are the warmest Romanian Carpathians region, due to the high frequency of mid-to-low elevations (below 1,500 m) and the prevailing humid and warm air masses advection. The share of the 'warm' belt in their overall surface area is above 50 % (Fig. 7.4). The 'moderately cool' and 'moderately warm' belts are best represented in the Eastern and Southern Carpathians, respectively (Table 7.4).

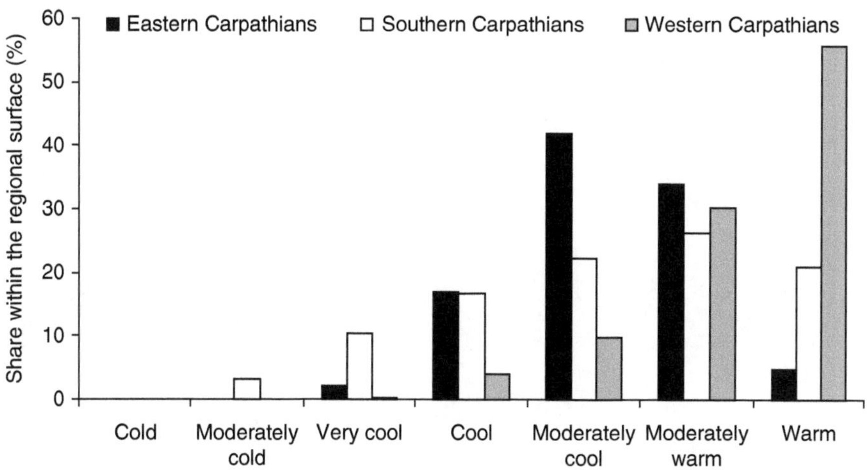


**Fig. 7.4** Spatial share of thermal belts in the Romanian Carpathians

**Table 7.4** Vertical thermal zonation of the Romanian Carpathians

| | | Romanian Carpathians branches | | |
|---|---|---|---|---|
| Thermal belts | Average annual temperature (°C) | Eastern Carpathians | Southern Carpathians | Western Carpathians |
| Cold | <−2 | n/a | 2,143–2,273 | n/a |
| Moderately cold | −2 ... 0 | 1,727–1,995 | 1,707–2,228 | n/a |
| Very cool | 0 ... 2 | 1,340–1,839 | 1,417–2,002 | 1,507–1,749 |
| Cool | 2 ... 4 | 949–1,669 | 1,178–1,736 | 1,115–1,624 |
| Moderately cool | 4 ... 6 | 672–1,357 | 874–1,489 | 857–1,372 |
| Moderately warm | 6 ... 8 | 379–1,070 | 585–1,191 | 508–1,134 |
| Warm | >8 | 131–751 | 182–833 | 64–826 |

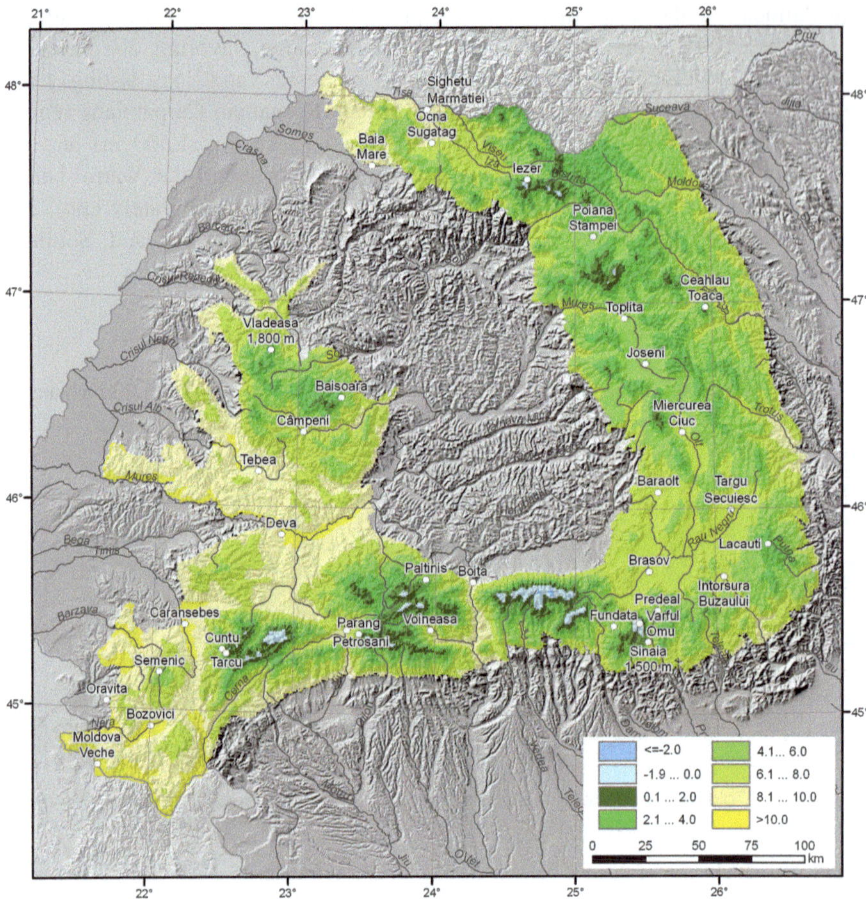


**Fig. 7.5** Annual $T_{avg}$ (°C) in the Romanian Carpathians

This zonation is reflected by the large differences of annual $T_{avg}$ between the foothills and the ridge areas of the Romanian Carpathians (Fig. 7.5.), of:

- 10.3 °C in the Southern Carpathians, between 7.9 °C at Petroşani (607 m) and −2.4 °C at Vf. Omu (2,504 m);
- 10.6 °C in the Western Carpathians, between 11.8 °C at Moldova Veche (82 m) and 1.2 °C at Vlădeasa 1,800 m (1,836 m);
- 8.0 °C in the Eastern Carpathians, between 8.7 °C at Sighetul Marmaţiei (275 m) and 0.7 °C at Ceahlău-Toaca (1,897 m).

The annual average temperature of the areas depicting the three vegetation belts covers a 5 °C range: from −0.9 °C in the alpine to 1.1 °C in the subalpine and 4.1 °C in the forest belts.

The position of some key isotherms explains the spread of the main vegetation formations and the typology of periglacial processes across the Romanian Carpathians, under the influence of altitude and latitude.

**Table 7.5** The elevation range (m) of some key $T_{avg}$ isotherms in the Romanian Carpathians

| Isotherms | Eastern Carpathians | Southern Carpathians | Western Carpathians |
|---|---|---|---|
| −2 °C | n/a | 2,143–2,229 | n/a |
| 0 °C | 1,581–1,902 | 1,645–2,055 | n/a |
| 2 °C | 1,310–1,682 | 1,371–1,782 | 1,507–1,637 |
| 3 °C | 1,271–1,620 | 1,331–1,515 | 1,137–1,511 |
| 10 °C (July) | 1,607 | 1,693 | 1,567 |

*The height of the 0 °C isotherm* ('freezing level') marks the areas in which the periglacial and frost processes prevails (Parish 2002). Investigating the favorability of ecological and bio-geographical conditions for high-altitude insects, Mani (1968) has illustrated the freezing level height in the most important mountain regions of the world. Relative to other mountain ranges (e.g. above 3,000 m in the Atlas, Caucasus, Tian-Shan and Altai Mountains, above 5,000 m in the Himalaya and Karakorum Mountains), the elevation of permanent air freezing in the Romanian Carpathians is one of the lowest. Yet, the freezing level height in the Romanian Carpathians is comparable to that in the European Alps and Pyrenees (at about 2,500 m), located in similar latitudinal conditions (between 42 and 49°N) and well above that in the Scandinavian Mountains (below 1,000 m), located at high latitudes. On average, the areas of prevailing periglacial and frost processes in the Romanian Carpathians are delineated by 1,764 m altitude in the Eastern Carpathians and 1,885 m in the Southern Carpathians (Table 7.5). The 0 °C isotherm can be considered a good proxy of the areas prone to long-lasting snow accumulation.

Hess (1965) found significant connections between the climatic snowline and *the −2 °C annual isotherm* across the Carpathians. In the Romanian Carpathians, this isotherm appears rather patchy, only in the highest massifs of the Southern Carpathians (the Făgăraş Mountains), at above 2,100–2,200 m. The areas delineated by this isotherm are prone to long periods of intense mechanical weathering (by freezing and freeze-thaw cycles), gravitational processes (creep, solifluction, falls, topples), moderate-to-low surface erosion and high intensity deflation and snow-driven processes.

The temperature conditions are the main limiting factors for the upper forest limit. The height of the *+2 °C annual isotherm* corresponds roughly to the upper tree line (timberline), delineating the areas of optimal climatic condition for forest growth. However, the upper boundary of forest vegetation is more clearly delineated in terms of thermal conditions in the areas of weak anthropogenic pressure and rather diffuses in those of extensive deforestation actions. On average, the isotherm is located at about 1,500 m in the Eastern Carpathians and 1,600 m in the Southern and Western Carpathians (Table 7.5). In addition, Urdea (2012) asserted that the 3 °C mean annual isotherm is an indicator of the lower limit of periglacial environments, which includes the *solifluction zone* (between the 3 °C and 0 °C isotherms), the *zone of complex periglacial processes* (between 0 °C and −3(−2)°C isotherms) and the *cryoplanation zone*, with intense weathering (below −3(−2)°C down to the −6 °C isotherm). This limit lowers to 1,300 m in the Southern and Western Carpathians and 1,450 m in the Eastern Carpathians.

Florescu (1981) indicated that the northern (latitudinal) border of forest vegetation in Europe delineates areas of daily average temperatures higher than 10 °C over at least 30 days. *The July isotherm of 10 °C* roughly corresponds to the upper forest line and the northern (latitudinal) tree line, as it is considered a proxy of heat deficiency during the vegetation growing season (EEA 2004). However, Landolt (1983) has revealed some exceptions for the 'rule of thumb' that the July isotherm of 10 °C has for some forest vegetation species (e.g. *Pinus cembra* in the Swiss Alps – 7.5 °C July isotherm). In the Romanian Carpathians, the July 10 °C isotherm is a reliable indicator of the vertical possible expansion of coniferous forests, only in the Western Carpathians, where the isotherm runs parallel with the upper tree line located at about 1,600 m according to CORINE LandCover dataset (2006). In the Eastern and Southern Carpathians, this isotherm runs below the upper tree line, which is found up to 1,750–1,800 m.

### 7.2.3 Annual Variation of Average Temperature

On average, air temperature values become negative from November at above 800 m and December, below 800 m, maintaining this characteristic until March, at above 800 m and February, below 800 m. The coldest month of the year is January at most weather stations and February, only at high-elevation sites of the Eastern and Southern Carpathians, generally located above 1,800 m. During the annual minimum, $T_{avg}$ reaches the lowest values in the Southern Carpathians, at about 2,500 m (−10.6 °C, Vf. Omu station) and the highest in the Western Carpathians, at about 1,400 m (−3.9 °C, Băişoara station) (Fig. 7.6). The range of the annual minimum in the areas higher than 800 m, is well below the freezing level as follows: −8.2°...−6.6 °C in the Eastern Carpathians, −10.6°...−4.3 °C in the Southern Carpathians and −7.2°...−3.9 °C in the Western Carpathians. Below 800 m, air temperature values in January are exclusively negative in the Eastern and Southern Carpathians (between −7.2 and −2.6 °C), and slightly above 0 °C, only in some areas of the Western Carpathians (e.g. 0.1 °C at the Moldova Veche station, in the Banat Mountains).

Air temperature becomes positive starting in April in most areas of the Romanian Carpathians, except high-elevation areas above 1,800 m (from May) and earlier, from March or even February, generally below 600–700 m.

The warmest month of the year is July at most weather stations and August only at some high-elevation sites located above 1,700–1,800 m in the Eastern and Western Carpathians (Lăcăuţi and Vlădeasa 1,800 m) and above 2,200 m in the Southern Carpathians (Vf. Omu and Ţarcu). Regionally, in the areas above 800 m, $T_{avg}$ ranges between 9.8 and 14.6 °C in the Eastern Carpathians, 5.9° and 14.8 °C in the Southern Carpathians and 9.9° and 14.0 °C in Western Carpathians (Fig. 7.6).

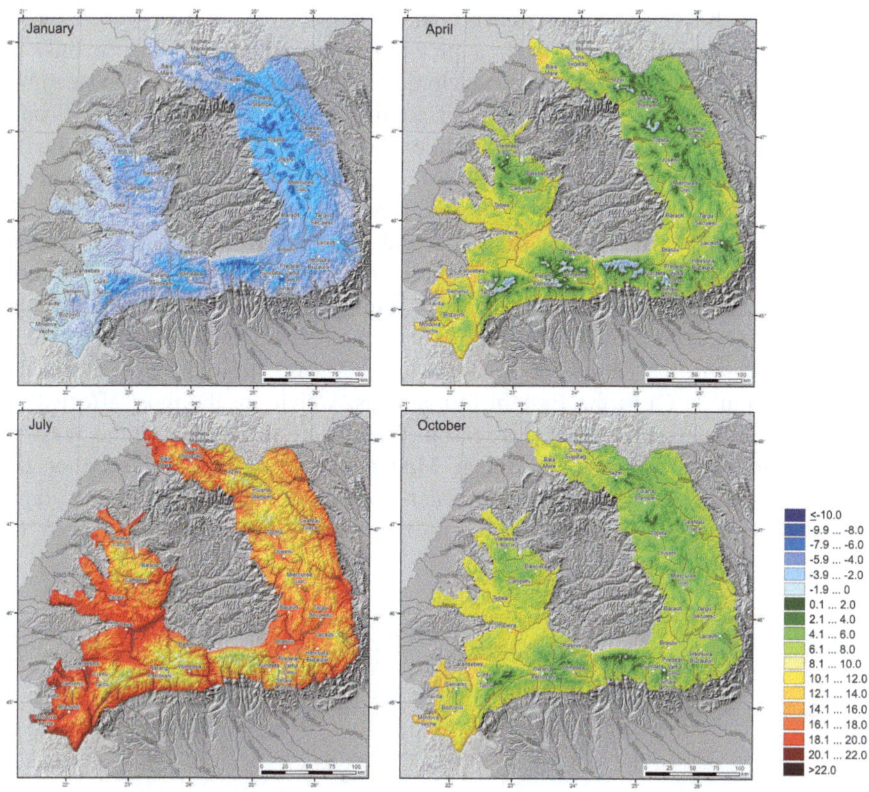


**Fig. 7.6** Monthly distribution of $T_{avg}$ (°C) across the Romanian Carpathians

The variation of $T_{avg}$ values at high-elevations (above 1,800 m) is limited to the 6–10 °C range. In depression areas (below 800 m), temperature values in July vary on average between 16 and 19 °C, exceeding 20–22 °C only at some low-elevation sites located in the Western Carpathians (the Banat Mountains).

$T_{avg}$ varies in the negative range from October only in the high alpine areas of the Southern Carpathians (above 2,500 m), November in most areas above 800 m of the Eastern Carpathians and December in the rest of the region, including the areas below 1,600 m in the Southern Carpathians, 1,400 m in the Western Carpathians and 700 m in the Eastern Carpathians.

The thermal differences between uphill (mountains) and foothill areas (depressions) over the year across the Romanian Carpathians are the largest in July-August, up to 9–12 °C, when $T_{avg}$ has the greatest decreasing rates with height, and smallest in January, down to 5–10 °C, as a result of frequent and persistent temperature inversions, which perturb significantly the vertical thermal zonation.

Seasonally, there is a close relationship between regional air temperature and large-scale flow patterns over these mountains. The temperature distribution shows similar patterns in all seasons.

In winter, the temperature regime is controlled by the cold northerly and easterly airflows (polar and arctic), under a prevailing anticyclonic regime (mainly in January and February) and weaker cyclonic activity over the October-December interval. The high frequency of cold-air pooling in this season, as result of both dynamical and radiative processes, might explain some local and regional patterns of $T_{avg}$ variation. Winters are up to 3 °C colder in the Eastern and Southern Carpathians than in the Western Carpathians, particularly in some northern massifs exposed to cold airflows of continental and maritime origins (e.g. the Rodna, Călimani, Ceahlău Mountains) and in some southern massifs, with a wide elevation range (e.g. the Făgăraş Mountains) (Figs. 7.7 and 7.8). By height, winters across the Romanian Carpathians are 3–7 °C colder in high-elevation areas above 1,700–1,800 m than in the depression areas (below 800 m). These differences are the lowest (about 3 °C) in the Eastern Carpathians, which are the most affected by temperature inversions during this season. Winters are the mildest in the Western Carpathians (the Apuseni and Banat Mountains), under the influence of the prevailing humid westerly and southwesterly airflows, respectively. $T_{avg}$ at above 800 m during the cold season shows narrow ranges in the Eastern (−7.6°...−5.5 °C)

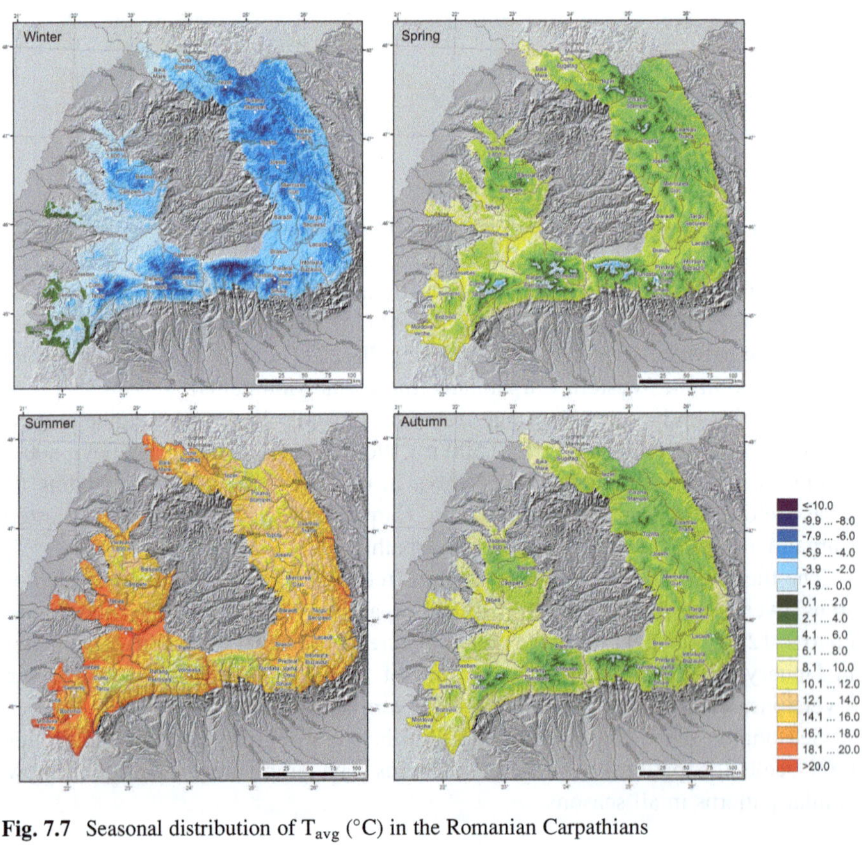


**Fig. 7.7** Seasonal distribution of $T_{avg}$ (°C) in the Romanian Carpathians

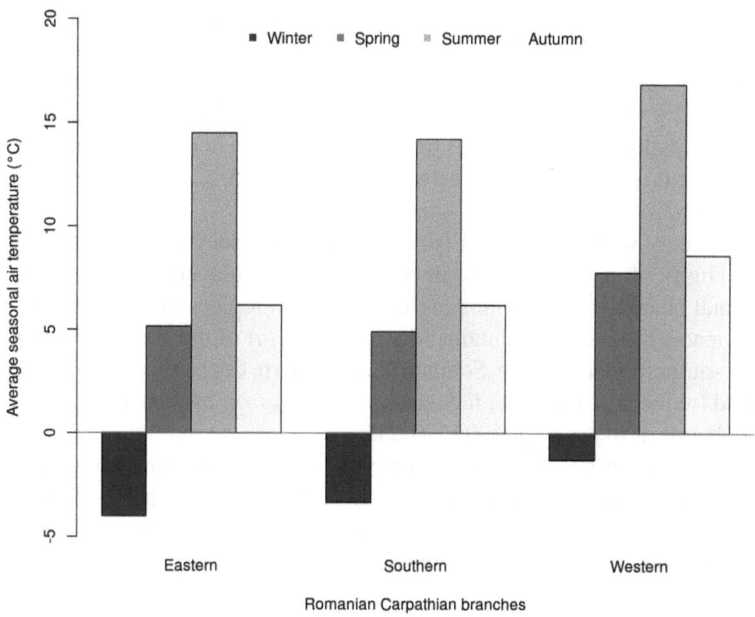


**Fig. 7.8** Regional averages of seasonal $T_{avg}$ values

and Western Carpathians (−6.7°...−3.4 °C), compared to the broader one in the Southern Carpathians (−9.8°...−3.5 °C), resulting from their largest vertical expansion. In this season, the coldest depressions are those located in the Eastern Carpathians (the Braşov, Ciuc, Întorsura Buzăului depressions), which are the most favorable areas for the formation of persistent cold-air pools and cold air downslopes drainage. Mihai (1975) showed that these areas are prone to intense temperature inversions, the inversion layer thickness extending up to 850 m. In these areas the $T_{avg}$ varies between −5.9 and −1.9 °C. The opposite situation is characteristic of the depressions located in the Western Carpathians, where $T_{avg}$ seasonal variation is only slightly negative, from −2.5 to −0.7 °C. Locally, in some of these areas (e.g. Caransebeş, Moldova Veche), the wintertime $T_{avg}$ stays positive (0.3–1.5 °C).

Summers are slightly cooler in the Eastern Carpathians (up to 1 °C cooler) than in the other Carpathian branches, only in the areas higher than 800 m (Fig. 7.7). This is due to their northern position, their great exposure to frequent passages of cold (polar or arctic) air masses, as well as to their dense forest cover. These atmospheric circulation patterns also explain the shallower TLRs in summer, particularly in the northern part of this Carpathian region. In the Eastern Carpathians, $T_{avg}$ values range between 9.1 and 13.9 °C at above 800 m and between 15.4 and 18.4 °C below 800 m. For comparison, the summer temperatures in the Southern Carpathians do not exceed 17 °C in depressions and fall below 5 °C in high alpine areas above 2,500 m.

The warmest summers are characteristic of the Western Carpathians, regardless of the elevation, under the dominant influence of westerly and south-westerly airflows (Fig. 7.8). This region is up to 1 °C warmer in the areas higher than 800 m and 2.4 °C, in the depressions, than the other Carpathian regions. In the Western Carpathians, the summertime $T_{avg}$ values range between 9.2 and 13.3 °C in mountain areas above 800 m and between 16.5 and 21.5 °C in low depression areas below 800 m.

Advection of warm tropical air from the Sahara (under the activity of the North-African High), often affects the southern part of Romania, and largely accounts for the thermal character of summers in the Carpathian region. Such advection particularly impacts the Banat Mountains (the southern part of the Western Carpathians) and the southern sides of the Southern and Eastern Carpathians, determining an increased frequency of summer hot days, when $T_{max}$ may exceed 30 °C, particularly in areas below 1,500 m (e.g. 2007, 2012).

Autumn and spring show fairly comparable $T_{avg}$ variation ranges (Fig. 7.8). In spring, temperature records are slightly negative at high-elevations (above 1,700 m) and positive in the rest of the region, whereas in autumn, they remain entirely positive, regardless of the elevation, all over the region (Fig. 7.7).

### 7.2.4 *Lapse Rates of Minimum ($TLR_{min}$) and Maximum Temperatures ($TLR_{max}$)*

A differentiation of temperature lapse rates into $TLR_{max}$ and $TLR_{min}$ was introduced in the study to highlight the particularities of the distribution of maximum and minimum temperatures with respect to elevation across the Romanian Carpathians. The elevation predictor explains about 90 % of the spatial distribution of extreme temperatures in the region, at both annual and seasonal time-scales.

The average regional lapse rates for extreme temperatures are generally steeper for maximum temperatures (about 0.74 °C/100 m) and shallower for minimum temperatures (about 0.46 °C/100 m). Regionally, the variations in the $TLR_{ext}$ are explained by the characteristics of the dynamical processes over the year in relation to the elevation range. The temperature inversions, cold-air pooling and downslope drainage during winter are likely to explain the reduced gradients of both minimum and maximum temperatures, especially in the Eastern Carpathians (Table 7.6). The vertical change rates of the maximum temperature are generally closer to the dry adiabatic lapse rates as Rolland (2003) emphasized, indicating stronger thermal contrasts between Western Carpathians (the warmest) and Eastern and Southern Carpathians (the coldest), particularly in summer, spring and autumn.

The extreme temperature datasets indicate consistent seasonal cycles of the $TLR_{ext}$, which are common with the $TLR_{avg}$, showing similar regional patterns between the three Carpathian branches. The lapse rates are smallest (0.34–0.46 °C/100 m) in winter and autumn minimum temperatures and largest (0.79–0.96 °C/100 m)

Table 7.6 Annual and seasonal lapse rates of the minimum and maximum temperatures in the Romanian Carpathians (°C/100 m)

| TLRs | Eastern Carpathians | Southern Carpathians | Western Carpathians |
|---|---|---|---|
| $TLR_{min}$ | | | |
| Annual | 0.41 | 0.49 | 0.50 |
| DJF | 0.34 | 0.44 | 0.43 |
| MAM | 0.49 | 0.57 | 0.55 |
| JJA | 0.46 | 0.53 | 0.55 |
| SON | 0.37 | 0.44 | 0.46 |
| $TLR_{max}$ | | | |
| Annual | 0.73 | 0.80 | 0.69 |
| DJF | 0.46 | 0.57 | 0.47 |
| MAM | 0.86 | 0.92 | 0.79 |
| JJA | 0.91 | 0.96 | 0.89 |
| SON | 0.70 | 0.76 | 0.67 |

in summer and spring maximum temperatures (Fig. 7.9). Differences in lapse rates have been found to be more substantial for maximum temperatures, indicating larger vertical changes with height, which can be attributed to the strong daytime sunshine warming on mountain slopes. Regionally, these differences are highly visible in the Southern Carpathians (the highest), relative to the other Carpathian regions.

### 7.2.5 Spatial Distribution of Minimum and Maximum Temperatures

The annual $T_{max}$ is positive over the entire Romanian Carpathians, varying from 4.3 to 10.6 °C in the mountain areas above 800 m and from 11.6 to 17.1 °C in depression areas (Fig. 7.10). Regionally, the differences between the foothills and ridge areas are substantial in terms of annual averages of $T_{max}$, reaching:

- 13.3 °C in the Southern Carpathians, between 0.7 °C at Vf. Omu (2,504 m) and 14.1 °C at Petroşani (607 m);
- 12.8 °C in the Western Carpathians, between 4.3 °C at Vlădeasa 1,800 m (1,836 m) and 17.1 °C at Moldova Veche (82 m);
- 10.3 °C in the Eastern Carpathians, between 4.3 °C at Ceahlău-Toaca (1,897 m) and 14.6 °C at Sighetul Marmaţiei (275 m).

The $T_{min}$ shows a larger variations across the Carpathian region on an annual average basis, from negative values of −5.0°...−0.1 °C in the high-elevation areas above 1,800 m and some depressions of the Eastern Carpathians, to positive values ranging from 0.6 to 4.0 °C in the depressions of the Eastern and Southern Carpathians, to 5.0–7.5 °C in some depressions of the Western Carpathians (generally, below 300 m) (Fig. 7.11). The largest differences between the foothills

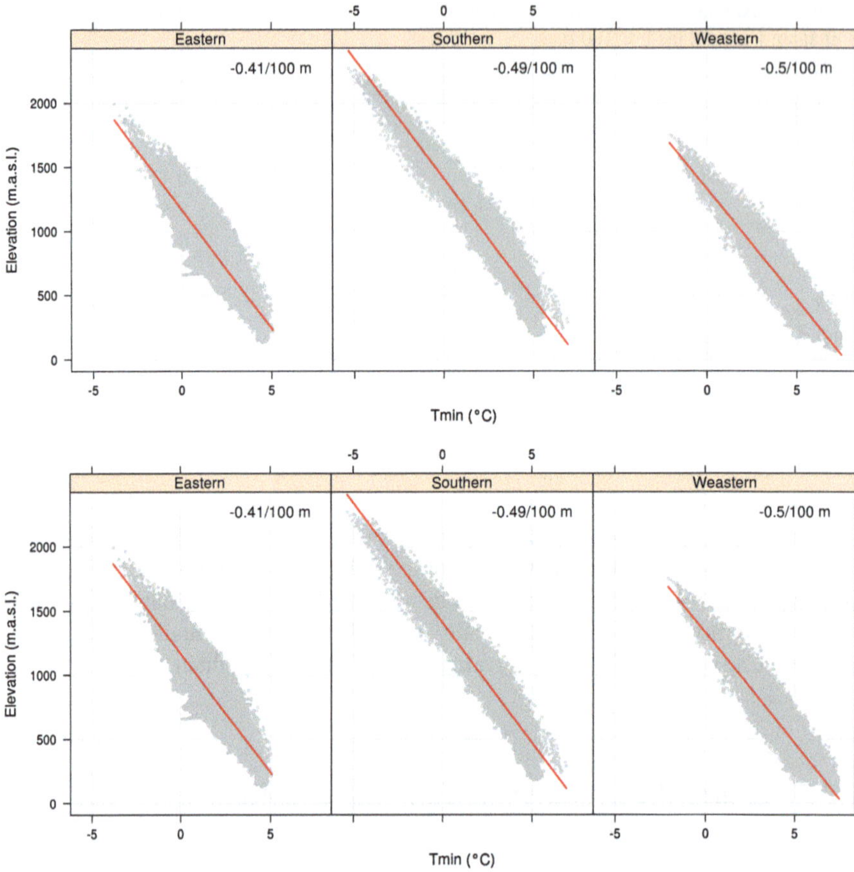


**Fig. 7.9** Lapse rates (°C/100 m) of the annual averages of $T_{min}$ and $T_{max}$ in the Romanian Carpathians

and high mountain peaks are met in the Western Carpathians, where they are about 9 °C. The average height of the freezing level derived from $T_{min}$ values is located at about 1,150 m in the Eastern Carpathians, 1,400 m in the Southern Carpathians and 1,350 m in the Western Carpathians.

### *7.2.6 Annual Variation of Minimum and Maximum Temperatures*

Barry (2008) showed that, while $T_{max}$ displays a closer relationship to global radiation totals, $T_{min}$ seems to be much more site-dependent, especially with respect to cold air drainage and pooling.

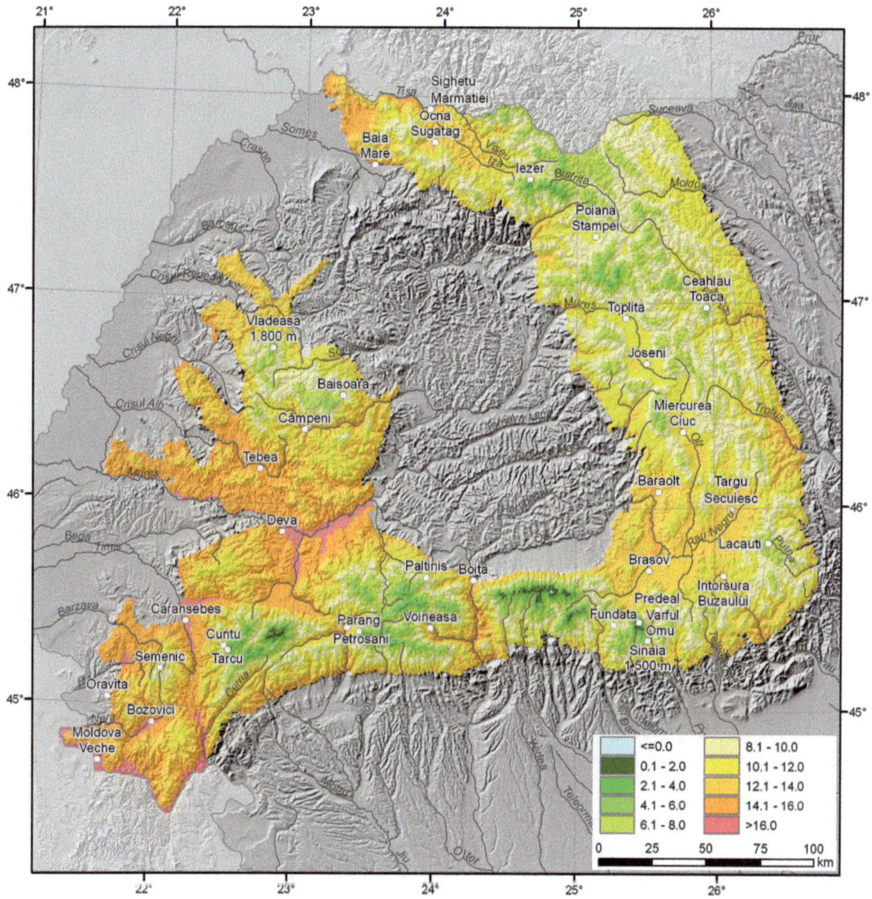


**Fig. 7.10** Mean annual $T_{max}$ (°C) in the Romanian Carpathians

In their monthly variation, the peak values of the $T_{min}$ and $T_{max}$ are recorded during the July–August interval. During the annual maximum, the peak values range between 11–16 and 25–28 °C, respectively in the low depression areas and between 9–14 and 3–7 °C, respectively, in those higher than 1,800 m.

During the annual minimum (January-February), $T_{max}$ values do not fall significantly below −8.0 °C not even in the high alpine areas, whereas $T_{min}$ may drop below −10 °C in most areas above 600 m of the Eastern Carpathians and 2,000 m of the Southern Carpathians. The Western Carpathians maintain the warmest, with slightly negative $T_{max}$ values (−4.5 to −0.1 °C) only in areas higher than 1,400 m and $T_{min}$ of −9.7 to −1.2 °C.

Seasonally, there is a region-wide hierarchization of extreme temperatures from the coldest to warmest: winter – spring – autumn – summer (Figs. 7.12 and 7.13). Winters display negative $T_{max}$ at 12 out of the 15 selected mountain sites located

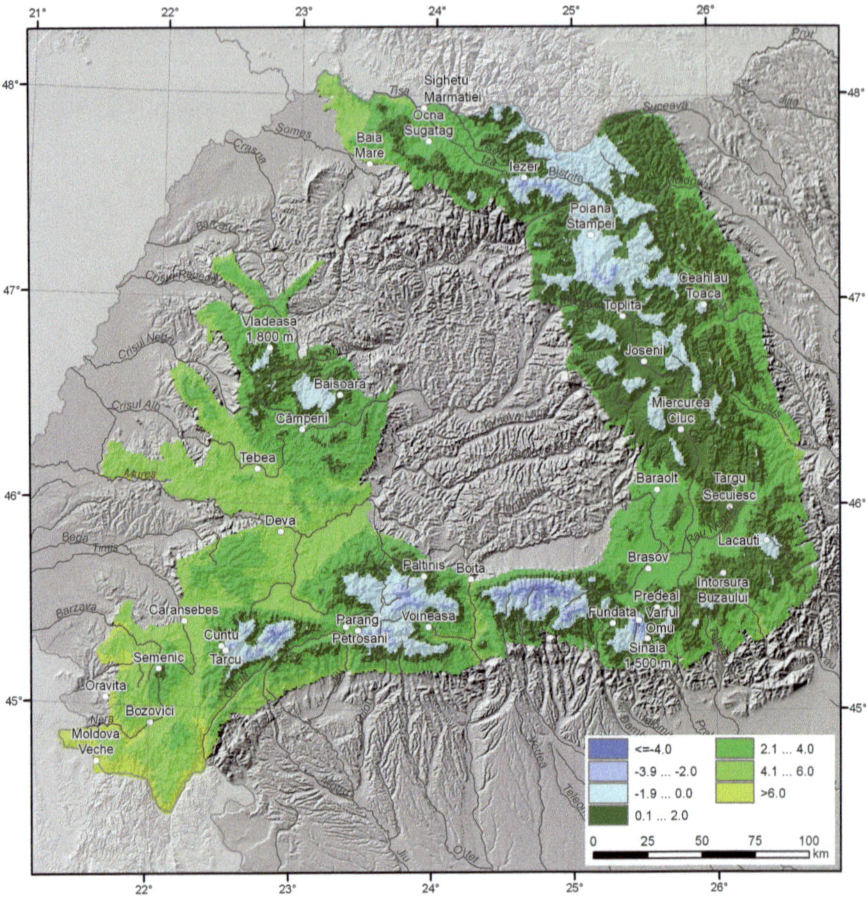


**Fig. 7.11** Mean annual $T_{min}$ (°C) in the Romanian Carpathians

above 800 m in the Romanian Carpathians, where values are generally 0.4–6.8 °C below the freezing level. In terms of $T_{min}$, winters are visibly colder at above 1,800 m, where averages drop 9–13 °C below the 0 °C threshold, and are generally cooler below this elevation, with average values of up to 10 °C below 0 °C. Summers are rather cool for $T_{max}$ in the Southern Carpathians and the averages do not exceed 20 °C in the areas above 800 m and 25 °C in those below this elevation. The Western Carpathians are prone to warmest summers. In terms of $T_{min}$, summers are rather cool in the Eastern and Southern Carpathians (2.8–12.6 °C) and may exceed 13 °C on average only in some depression areas of the Western Carpathians (below 300 m). The regional patterns of extreme temperatures in spring and autumn are fairly comparable across the Romanian Carpathians (Figs. 7.12 and 7.13).

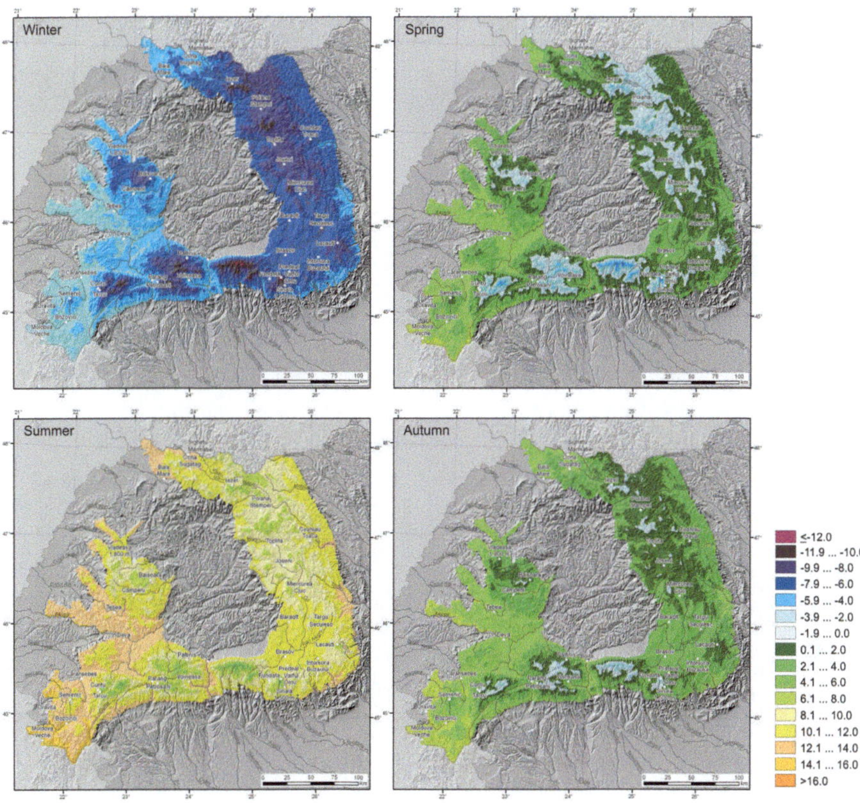


**Fig. 7.12** Mean seasonal $T_{min}$ (°C) in the Romanian Carpathians

### *7.2.7 Extreme Temperature Records*

The probability density function (PDF) of daily $T_{max}$ values at the selected mountain stations show a limited heat stress exposure in the region (Fig. 7.14). Most $T_{max}$ values lie between −5 and +25 °C, with a 0.9 probability across the entire region. The upper and lower bounds of regional $T_{max}$ PDF depict the threshold beyond which $T_{max}$ enters into the 'extreme' domain and they are fairly similar to the 90th and 10th quantiles. The upper bounds are generally marked by $T_{max}$ values of 25 °C and regionally their occurrence probability is one of the lowest (below 0.1). Yet, on a site scale, this bound might be exceeded with less than 0.05 probability. Such values are specific to mid-elevation areas (below 1,500–1,600 m), which are often subject to warm and humid airflows from Southern Europe, local warming in summer due to urbanization or foehn processes (e.g. on the eastern side of the Apuseni Mountains – Western Carpathians). The lower bounds of $T_{max}$ PDF across the Romanian Carpathians are usually marked by the −10 °C threshold (below 0.005 probability).

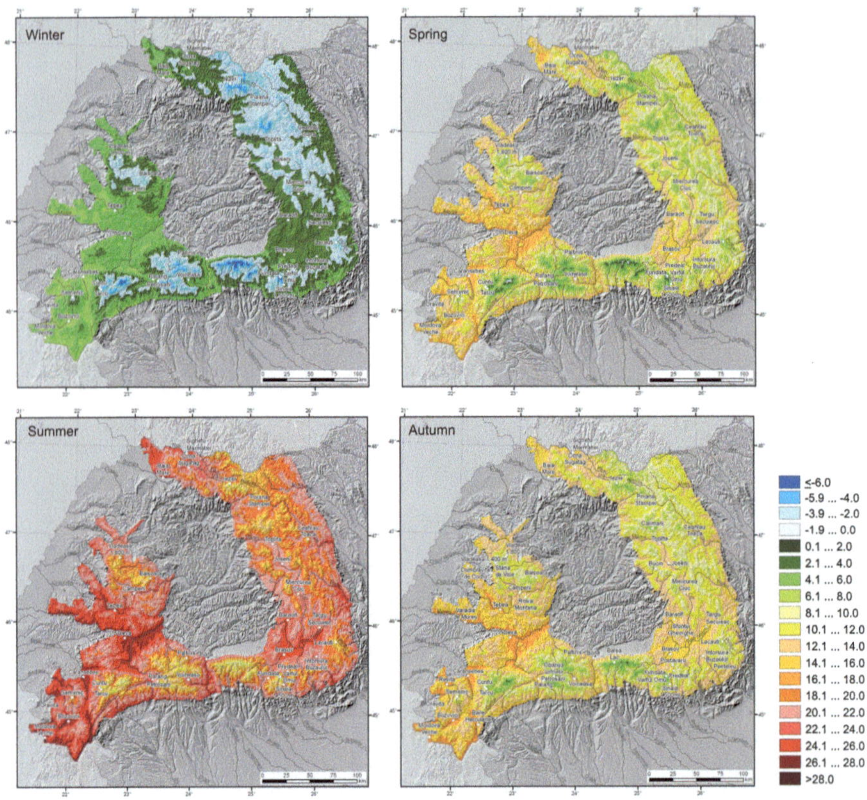


**Fig. 7.13** Mean seasonal $T_{max}$ (°C) in the Romanian Carpathians

The $T_{max}$ absolute maxima registered since the beginning of instrumental meteorological records in areas higher than 800 m had rarely reached or exceed the 30 °C threshold, which depicts the occurrence of a tropical day. Such values were specifically recorded sparsely up to 1,300–1,500 m in the Southern and Western Carpathians (at Predeal, Sinaia 1,500 m, Păltiniş, Predeal and Băişoara sites) and up to 1,000 m in the Eastern Carpathians (Poiana Stampei). The $T_{max}$ historical records in the areas above 800 m range between 22 and 33 °C. The absolute maximum measured in such areas reached 33.2 °C on July 5th, 2000, at Predeal station (1,090 m), in the Southern Carpathians.

The $T_{max}$ absolute records in the depression areas of the Romanian Carpathians exceeded 39 °C: 39.4 °C in the Eastern Carpathians (Sighetu Marmaţiei, 275 m), 40.0 °C in the Southern Carpathians (Deva, 240 m), reaching the highest value of 44.0 °C in the Western Carpathians, at Moldova Veche station (82 m), on July 24, 2007 (Table 7.7). This value was only 0.5 °C below the national record (Ion Sion 44.5 °C/August 10, 1951 in the Bărăgan Plain – southern Romania).

The observed temperature rise in the last two decades, particularly in summer, when heat waves became more frequent and more intense, played a key role in

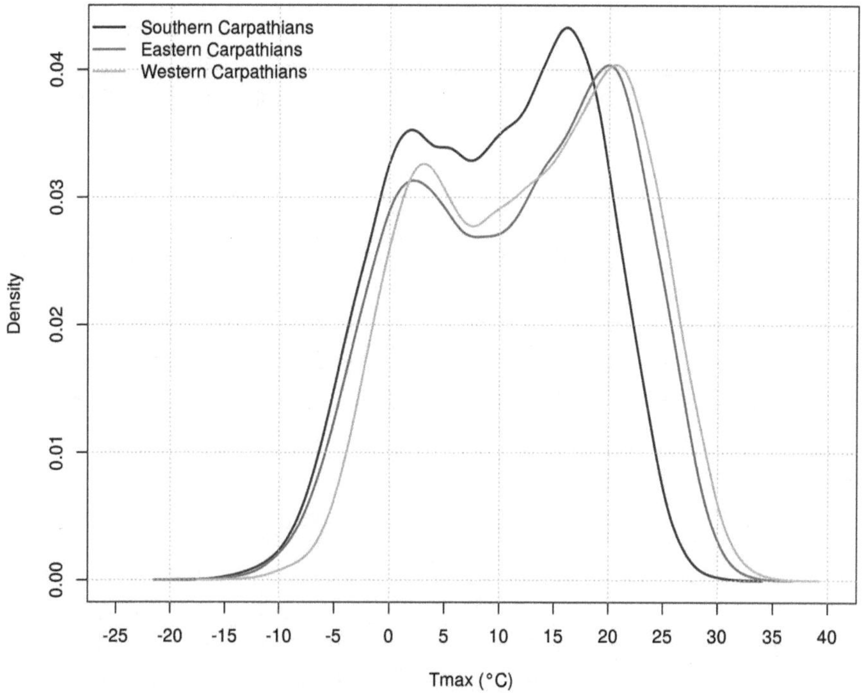


**Fig. 7.14** A comparison of regional $T_{max}$ PDFs in the Romanian Carpathians

explaining the occurrence of $T_{max}$ absolute records in many Carpathian areas. The $T_{max}$ absolute records were dominantly recorded after 2000 across the Romanian Carpathians region, accounting for 63 % of the total number of cases, suggesting a shift towards a warming climate in the region. The summer months of 2000, 2007 and 2010 hold 64 % of the total $T_{max}$ records at above 800 m and up to 54 % of those in areas below 800 m. These months were particularly warmer than normal also on a national scale due to successive and intense heat spells (Bogdan et al. 2011; Annual Bulletin on the Climate in WMO Region VI – Europe and Middle East 2000, 2007, 2010).

The PDFs of daily $T_{min}$ values recorded above 1,000 m altitude reveal a moderate exposure to freezing across the region (Fig. 7.15), as the lower bounds depicting values below −10 °C have a rather low occurrence probability (below 0.1), from 0.05 % (Western Carpathians) to 0.09 % (Eastern Carpathians). Most $T_{min}$ values lie between −5.0 and +10.0 °C as in the case of $T_{max}$, with a 0.63 probability at all mountain sites.

The $T_{min}$ values in the regional PDF reveal a negatively skewed distribution. The upper bounds, depicting values higher than 15 °C, have rather low probabilities across the region, especially in the areas below 1,500 m of the Southern and Western Carpathians. In a daily variation, $T_{min}$ records of 25° below the freezing level and 20 °C above it are quite exceptional across the region, underlying the

**Table 7.7** Maximum and minimum absolute temperature values measured in the Romanian Carpathians until 2010 relative to the national records

| | Romania | | Eastern Carpathians | | Southern Carpathians | | Western Carpathians | |
|---|---|---|---|---|---|---|---|---|
| Month | Absolute $T_{max}$ | Absolute $T_{min}$ | Absolute $T_{max}$ | Absolute $T_{min}$ | Absolute $T_{max}$ | Absolute $T_{min}$ | Absolute $T_{max}$ | Absolute $T_{min}$ |
| January | 22.0 °C/ Jan.7, 2001 (Oravița) | −38.5 °C/Jan.25, 1942 (Bod) | 18.0 °C/Jan.16, 1947 (Brașov) | −38.4 °C/ Jan.23, 1985 (Miercurea Ciuc) | 17.8 °C/ Jan.7, 2001 (Petroșani) | −34.4 °C/ Jan.3, 1979 (Țarcu) | 22.2 °C/ Jan.7, 2001, (Oravița) | −32.7 °C/ Jan.24, 1963 (Câmpeni) |
| February | 26.0 °C/ Feb.27, 1995 (Medgidia) | −38.0 °C/Feb.10, 1929 (Vf. Omu) | 18.8 °C/Feb.28, 1994 (Baraolt) | −35.8 °C/ Feb.8, 2005 (Întorsura Buzăului) | 20.8 °C/ Feb.23, 1977 (Deva) | −38.0 °C/ Feb.10, 1929 (Vf. Omu) | 23.3 °C/ Feb.22, 1966 (Moldova Veche) | −32.2 °C/ Feb.11, 1929 (Caransebeș) |
| March | 32.8 °C/ Mar.30, 1952 (Odobești) | −31.4 °C/Mar.9, 1952 (Întorsura Buzăului) | 27.8 °C/Mar.31, 1975 (Baraolt) | −31.4 °C/ Mar.9, 1952 (Întorsura Buzăului) | 29.9 °C/ Mar.30, 1952 (Boița) | −29.6 °C/ Mar.5, 1987 (Vf. Omu) | 28.7 °C/ Mar.24, 1977 (Mol-dova Veche) | −27.2 °C/ Mar.5, 1987 (Bozovici) |
| April | 35.5 °C/ Apr.10, 1985 (Bechet) | −26.0 °C/Apr.1,3,4,8,15, 1940 (Vf. Omu); −14.4 °C/ Apr.7, 1978 (Întorsura Buzăului) | 34.0 °C/Apr.22, 1950 (Sighetul Marmației) | −17.4 °C/ Apr.7, 1978 (Iezer) | 32.3 °C/ Apr.17, 1956 (Deva) | −26.0 °C/ Apr.1,3, 4,8,15, 1940 (Vf. Omu) | 30.6 °C/ Apr.24, 1968 (Țebea) | −16.2 °C/ Apr.7, 2003 (Vlădeasa 1,800 m) |
| May | 40.8 °C/ May27, 1950 (Mărculești) | −16.0 °C/May14,15, 1940 (Vf. Omu); −9.6 °C/May1, 1940 (Câmpulung Moldovenesc) | 33.0 °C/May27, 2003 (Sighetul Marmației) and May14, 1958 (Joseni) | −11.8 °C/ May2, 2007 (Ceahlău-Toaca) | 34.6 °C/ May26,27, 1950 (Deva) | −16.0 °C/ May14, 15, 1940 (Vf. Omu) | 35.1 °C/ May16, 1969 (Mol-dova Veche) | −9.6 °C/ May13, 1978 (Vlădeasa 1,800 m) |
| June | 42.0 °C/ Jun.29, 1938 (Oravița) | −12.0 °C/Jun.5,6, 1939 (Vf. Omu); −2.7 °C/ Jun.13, 1950 (Întorsura Buzăului) | 35.2 °C/Jun.18, 1968 (Sighetul Marmației) | −5.9 °C/ Jun.7, 1962 (Lăcăuți) | 35.6 °C/ Jun.19, 30, 1952, 1963 (Deva) | −12.0 °C/ Jun.5,6, 1939 (Vf. Omu) | 42.0 °C/ Jun.29, 1938 (Oravița) | −7.6 °C/ Jun.7, 1962 (Vlădeasa 1,800 m) |

| | | | | | | | | |
|---|---|---|---|---|---|---|---|---|
| July | 43.5 °C/ Jul.5, 2000 (Giurgiu) | −8.0 °C/Jul.6, 1933 (Vf. Omu); −1.3 °C/Jul.18, 1989 (Poiana Stampei) | 38.5 °C/Jul.23, 1939 (Ocna Şugatag) | −2.5 °C/ Jul.5, 1984 (Ceahlău-Toaca) | 40.0 °C/ Jul.24, 2007 (Deva) | −8.0 °C/ Jul.6, 1933 (Vf. Omu) | 44.0 °C/ Jul.24, 2007 (Moldova Veche) | −2.7 °C/ Jul.22, 1978 (Vlădeasa 1,800 m) |
| August | 44.5 °C/ Aug.10, 1951 (Ion-Sion) | −7.0 °C/Aug.20, 1940 (Vf. Omu); −1.9 °C/ Aug.27, 1980 (Miercurea Ciuc) | 39.4 °C/Aug.22, 1992 (Sighetul Marmaţiei) | −3.3 °C/ Aug.28, 1984 (Ceahlău-Toaca) | 39.7 °C/ Aug.16, 1952 (Deva) | −7.0 °C/ Aug.20, 1949 (Vf. Omu) | 41.0 °C/ Aug.11, 1994 (Moldova Veche) | −2.4 °C/ Aug.26, 1980 (Vlădeasa 1,800 m) |
| September | 43.5 °C/ Sep.8, 1946 (Strehaia) | −15.0 °C/Sep.18, 1935 (Vf. Omu); −10.7 °C/ Sep.21, 1977 (Topliţa) | 37.0 °C/Sep.9, 1946 (Braşov) | −10.7 °C/ Sep.21, 1977 (Topliţa) | 38.2 °C/ Sep.7, 1946 (Deva) | −15.0 °C/ Sep.18, 1935 (Vf. Omu) | 39.0 °C/ Sep.7, 1946 (Caransebeş) | −8.8 °C/ Sep.29, 1970 (Vlădeasa 1,800 m) |
| October | 39.0 °C/ Oct.3, 1952 (Armăşeşti) | −21.3 °C/Oct.27, 1988 (Întorsura Buzăului); −17.0 °C/Oct.31, 1920 (Roman) | 33.0 °C/Oct.2, 1952 (Braşov) | −21.3 °C/ Oct.27, 1988 (Întorsura Buzăului) | 30.0 °C/ Oct.1, 1991 (Deva) | −19.0 °C/ Oct.26, 1946 (Vf. Omu) | 34.0 °C/ Oct.6, 1935 (Caransebeş) | −14.4 °C/ Oct.29, 1997 (Vlădeasa 1,800 m) |
| November | 30.5 °C/ Nov.1, 1926 (Călăraşi) | −30.8 °C/Nov.30, 1957 (Vf. Omu); −28.2 °C/ Nov.27, 1993 (Joseni) | 26 9 °C/Nov.1, 1926 (Braşov) | −28.3 °C/ Nov.27 1993 (Joseni) | 27.4 °C/ Nov.1, 1926 (Deva) | −30.8 °C/ Nov.30, 1957 (Vf. Omu) | 37.4 °C/ Nov.1, 1926 (Deva) | −19.8 °C/ Nov.29, 1989 (Vlădeasa 1,800 m) |
| December | 23.4 °C/ Dec.5, 1985 (Câmpina) | −34.5 °C/Dec. 25, 1998 (Întorsura Buzăului) | 22.7 °C/Dec.9, 1960 (Întorsura Buzăului) | −34.5 °C/ Dec.25, 1998 (Întorsura Buzăului) | 19.8 °C/ Dec.14, 1957 (Deva) | −32.4 °C/ Dec.1, 1957 (Vf. Omu) | 22.7 °C/ Dec.9, 1960 (Întorsura Buzăului) | −24.6 °C/ Dec.15, 1961 (Vlădeasa 1,800 m) |

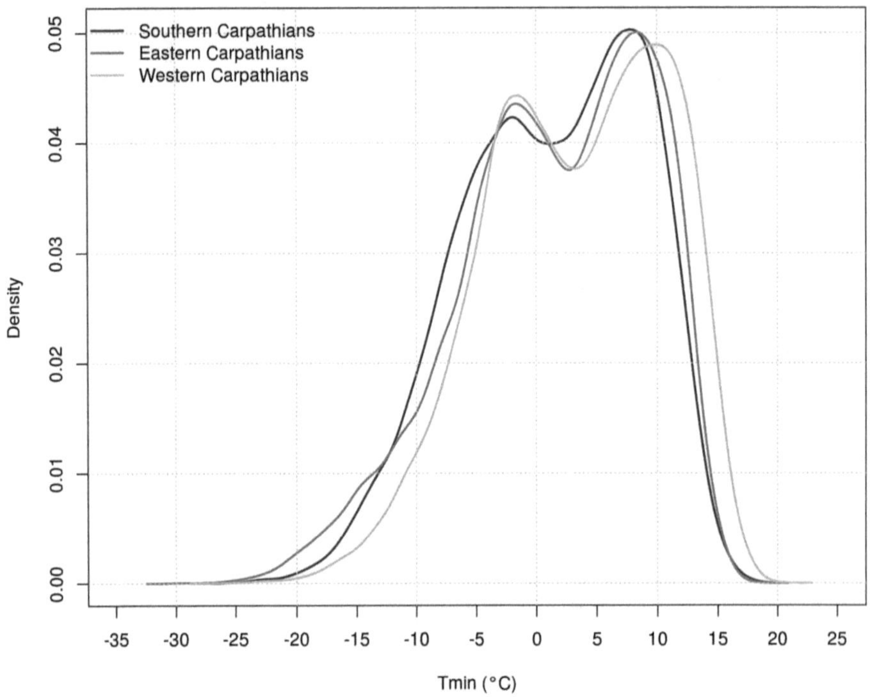


**Fig. 7.15** A comparison of regional $T_{min}$ PDFs in the Romanian Carpathians

extreme $T_{min}$ domain. The probability of air freezing ($T_{min} \leq 0$ °C) across the Romanian Carpathians is lower (15–20 %) than that of the freeze-free period ($T_{min} > 0$ °C). The frequency of freezing days across the Romanian Carpathians region ranges between 144 and 180 days in the forest areas (up to 1,600–1,700 m), 189–203 days in sub-alpine areas (1,700–1,900 m) and exceeds 200 days in the alpine ones (above 2,000 m). The intensity of air freezing across the Romanian Carpathians is moderate-to-severe, as its average intensities range between −15 and −10 °C (Fig. 7.16).

This phenomenon was severe in some outstanding cold winter months (e.g. January 1963, 1985; February 1929, 1985), when its intensity was more than 30 °C below the freezing level during some episodes of intense cooling. These months hold about 60 % of the total absolute minima registered across the study region. The absolute minima in the areas above 800 m reached −38.0 °C/February 10, 1929 at Vf. Omu station (2,504 m), in the Southern Carpathians, a value fairly comparable to the absolute minima in Romania (−38.5 °C/January 25, 1942 at Bod), which was registered in the Brașov Depression (the Eastern Carpathians).

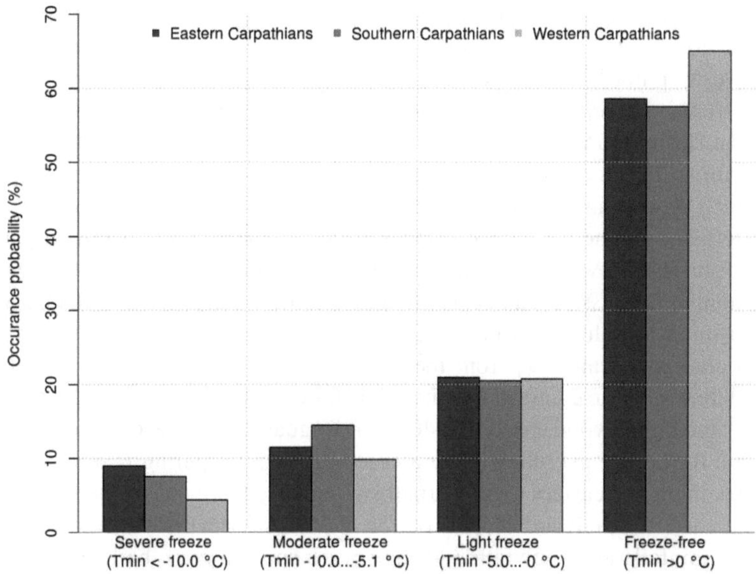


**Fig. 7.16** Regional freeze and freeze-free probabilities in the Romanian Carpathians

## 7.3 Precipitation

The distribution of precipitation amounts in the Romanian Carpathians is highly variable, strongly dependent on the underlying topography and orographic effect, in relation to regional flow dynamics. Beşleagă (1979) outlined the complexity of the precipitation genesis process and also the role of some other factors largely involved in this process (i.e. the upward and downward transfer of the water vapor flux, the moisture content of air masses advections and the vertical profile of air temperature, moisture and wind speed in the lower atmosphere). Bâzâc (1983) emphasized the importance of the Carpathian orographic effect on the spatial distribution of precipitation in Romania, which is also a measure of air masses instability potential and cloud base height, as well as of the direction and speed of the airstreams crossing the mountain barrier.

Due to their position within the continent, the precipitation regime of the Romanian Carpathians is subject to the influence of both important high and low pressure centers (see Chap. 6). Quantitatively, the maritime tropical and polar advective air streams, associated with cyclonic structures, are the main contributors to weather instability and shower activity in these mountains, as described below:

- *The Icelandic cyclones*, coming from the North Atlantic Ocean, are responsible for the occurrence of the primary (summer) precipitation peak of the year; the humid air advection at the southern edge of the Icelandic Low visibly increase weather instability in summer (frequent thunderstorms and heavy rainfall spells likely to trigger floods under large precipitation amounts) and produce wet weather in late autumn and early winter; such weather patterns are observed in most Carpathian areas, depending on slope exposure (particularly on the western, north-western and northern slopes);

- *The Mediterranean cyclones,* originating from the western and central Mediterranean Basin (more often than those with "normal" TransBalkanic trajectories[2]) determine the secondary precipitation peak of the year (late autumn and early winter), producing significant amounts and wind intensifications on the south-western and southern windward slopes of the Southern and Western Carpathians (the pressure lows describing north-eastern trajectories, which cross the Pannonian Depression), and exceptionally, on the external side of the Curvature Carpathians (the low pressures with eastern trajectories); when interacting with cold continental airflows from the east, they bring severe snowstorms (under great thermo-baric contrasts), with major snow accumulations (e.g. 40–50 mm in 24 h) and higher wind speed (30–40 m/s) (Bogdan 1978); the cyclonic structures with a retrograde evolution[3] play a crucial role in triggering severe weather in the southern and eastern extra-Carpathian regions, as well as on the southern and south-eastern windward slopes of the Eastern Carpathians, associated with heavy rainfalls and snowfalls (during severe snowstorms), hail, thunderstorms and gustiness.

### *7.3.1 Vertical Zonation and Spatial Distribution*

The effect of altitude on precipitation amount measured at the existing weather stations has been investigated in several studies focusing on different areas of the Romanian Carpathians (e.g. (Apăvăloaie 1972; Neamu and Teodoreanu 1972; Niculescu 1993; Voiculescu 2002; Gaceu 2005; Bogdan 2008). These works emphasize the role of the underlying (local) topography and prevailing atmospheric flow types in the process of precipitation generation and in explaining the patterns of spatial distribution of the precipitation amounts. Accordingly, there is a general linear increase of precipitation with elevation and the western and northern mountain slopes are accumulate the greatest amounts, as they are particularly subject to moist maritime and oceanic airflows (e.g. Dragotă 2006; Clima 2008).

Statistically, the altitude explains up to 58 % of the precipitation distribution in the Romanian Carpathians. The average precipitation gradient in this mountain range is 20.5 mm/100 m, less than that determined in the Polish Carpathians (60 mm/100 m) (Święchowicz 2012). Regionally, the precipitation gradient values

---

[2] Out of the six typical trajectories of these Mediterranean Cyclones towards Central and South-Eastern Europe, Şorodoc (1962) and Bordei-Ion (1983) identified four main trajectories with significant influence on weather types in Romania and in the Romanian Carpathians.

[3] The Mediterranean cyclones on eastward trajectories, reaching the Black Sea enhance their moisture content and evolve further on an opposite trajectory (i.e. most commonly westward and seldom northward).

are the steepest in the Western Carpathians (24.68 mm/100 m) and the shallowest one in the Southern Carpathians (15.75 mm/100 m). Seasonal precipitation patterns over the Romanian Carpathians, induced by the synoptic flow features (e.g. cyclones tracks and depth) in interaction with the orography, are well-reflected by the seasonal gradient values. Generally, the steepest gradients are characteristic of summer in the range of 9–4 mm/100 m and the least in winter, 1.5–2.5 mm/100 m (Table 7.8).

The vertical increase of precipitation is notably intensified by the upslope flow of air masses on the windward sides of mountain barriers up to a saturated air level (the condensation level – CL). Above the CL, precipitation gradient decreases. The height of the CL has been widely documented in several studies (e.g. Neamu and Teodoreanu 1972; Stoenescu 1951; Apăvăloaie 1972; Zăvoianu et al. 1995; Dragotă 2006; Bogdan 2008), but its position exhibits rather local validity, being strongly dependent on the measurement network density and the length of observation period. Table 7.9 summarizes the main findings of some representative contributions to the knowledge of the height of the mean CL in the three regions of the Romanian Carpathians.

**Table 7.8** Seasonal precipitation gradient (mm/100 m) in the Romanian Carpathians

| Carpathian branches | DJF | MAM | JJA | SON | Annual |
|---|---|---|---|---|---|
| Eastern Carpathians | 2.54 | 5.08 | 10.29 | 3.40 | 21.01 |
| Southern Carpathians | 1.93 | 3.54 | 8.73 | 2.81 | 15.75 |
| Western Carpathians | 1.52 | 5.81 | 13.88 | 3.19 | 24.68 |

**Table 7.9** The height of mean condensation level in the Romanian Carpathians

| Romanian Carpathians branches | Key findings in the specialist literature |
|---|---|
| Eastern Carpathians | 1,300–1,800 m (Apăvăloaie 1972); low elevation CL between 1,000–1,300 m and high elevation CL at above 1,700 m, on the north-western slope (Neamu and Teodoreanu 1972) |
| Southern Carpathians | "The optimum CL" in the Southern Carpathians is at about 1,600–1,800 m (Martonne 1907); the Bucegi Mountains: 1,600–1,800 m (Stoenescu 1951), 1,500–1,600 m (Bogdan 2008), 1,300 m in the area between the Făgăraş and the Bucegi Mountains (Neamu and Teodoreanu 1972); the Rucăr-Bran Corridor: 1,700–2,000 m (Teodoreanu 1980); the Făgpraş Mountains: "high-elevation CL" at about 2,000 m (Zăvoianu et al. 1995; Voiculescu 2002; Bogdan 2008); the Ţarcu-Făgăraş Mountains area: 1,400 m (Neamu and Teodoreanu 1972) |
| Western Carpathians | "The optimum CL" in the Western Carpathians is at 1,200–1,400 m (Martonne 1907); the Apuseni Mountains: 1,000–1,100 m (Gaceu 2005); 1,100 m (the lowest CL across the Romanian Carpathians) (Neamu and Teodoreanu 1972) |

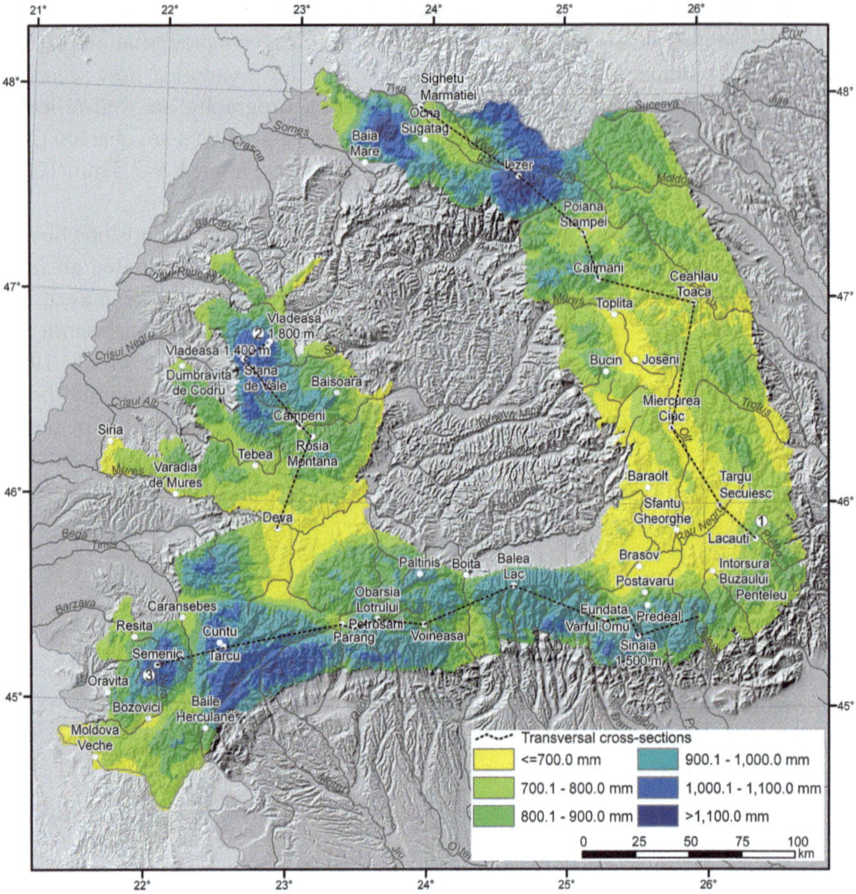


**Fig. 7.17** The distribution of the mean annual precipitation amounts (mm) in the Romanian Carpathians

The climate of the Romanian Carpathians is rather moderately humid, as over more than 70 % of their surface area the annual precipitation is below 900 mm. Annual precipitation totals increase from 500–800 mm in the foothill areas below 800 m to 900–1,300 mm in the areas above 800 m (Fig. 7.17). The Western and Southern Carpathians are the wettest, with a mean annual precipitation of about 800–900 mm, while the Eastern Carpathians are the driest (below 700 mm per year). The slopes of the western Apuseni and Banat Mountains (the Western Carpathians), Rodna and Maramureş Mountains (the Eastern Carpathians) and the southwestern ones of the Retezat Mountains (the Southern Carpathians) are particularly humid and they receive more than 1,000–1,100 mm of precipitation yearly. In these areas, the orographic effect accounts the most for the incoming moisture of maritime origin, enhancing (suppressing) the precipitation amounts on windward (leeward) slopes. Based on in situ measurements of precipitation amounts the

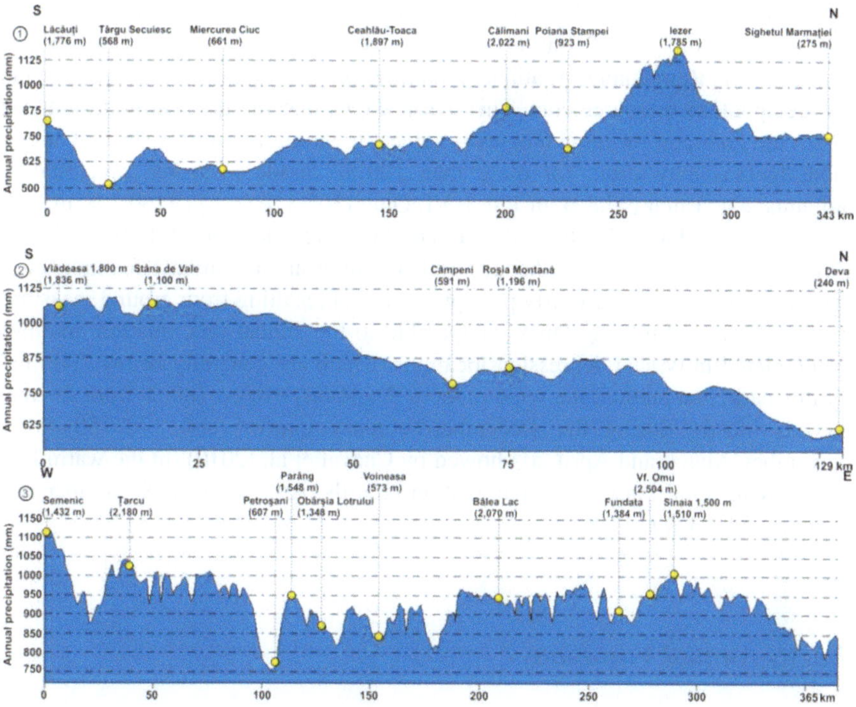


**Fig. 7.18** Cross-sections of the mean annual precipitation amounts in the Romanian Carpathians

windward side of the Apuseni Mountains is exceptionally wet, accumulating more than 1,600 mm per year (1,631.5 mm at Stâna de Vale station, located at 1,100 m a.s.l.) (Clima 2008). This area is also recognized as the wettest area in Romania (Gaceu 2005).

Generally, there is an eastward decrease of precipitation amounts across the Romanian Carpathians, as the moisture content of the westerly (oceanic) and southwesterly (Mediterranean) airflows gradually diminish (Fig. 7.18).

### 7.3.2 *Wind-Induced Bias*

High-elevation areas are particularly affected by wind-induced losses of precipitation amount, caused by the disturbance of the wind field around the gauge, evaporation and wetting losses of water in the funnel before it is measured, and splashing of rain drops upon impact (Sevruk 1982; Duchon and Essenberg 2001). The wind effect in complex orography terrains implies much underestimation of precipitation records, particularly of those resulting from solid fall accumulations.

An exhaustive analysis (Cheval et al. 2010) of errors in precipitation records over 1961–2009, found at 159 stations in Romania (including high-elevation stations located in the Romanian Carpathians), outlines the great perturbing effect of wind on precipitation measurements. This study shows that the greatest wind-induced losses are common for mountain areas subject to strong zonal flow. In such areas, monthly bias may count for more than 100 % of the measured precipitation amounts: e.g. Țarcu (37.8 % in June), Vf. Omu (107.3 % in April, 60.4 % in May and 39.4 % in June), Ceahlău-Toaca. The latter site (located at 1,897 m in the Eastern Carpathians), which displays rather similar alpine climatic features, is of special interest, since the yearly precipitation total resulting from ground measurements (about 670 mm) is fairly comparable to the annual precipitation amounts accumulated in typical depression locations, which are generally below 700 mm. At this particular station, the wind induces deviations exceed 180 % from the real records mostly during the December-February interval and 100–150 % in November, March and April, as showed by Cheval et al. (2010). In the warm half of the year, wind biases the precipitation records particularly in September and October (54–93 %).

### 7.3.3 Annual Variation

Studying the precipitation distribution types in Romania, Dragotă (2006) assigned to the Carpathian region '*the mountain distribution type*'. This distribution type is characterized by a main precipitation summer peak in June (23 sites) or July (9 sites) and is related to intense frontal activity and thermal convection processes. During the main pluviometric maximum, precipitation amounts between 105 and 160 mm in the areas above 800 m and between 78 and 110 mm in depressions. The summer maximum generally accounts for 12–16 % of the total annual precipitation. The second precipitation maximum occurs in most areas in the late autumn-to-early winter interval (in October, November, but particularly in December) and it is mainly of cyclonic origin, locally enhanced by the orographic effect. The accumulated precipitation during this maximum accounts for less than 8 % of the annual precipitation amount and does not exceed 100 mm.

The pluviometric minimum in the Romanian Carpathians is specific to the cold season months, when the anticyclonic regime prevails (typically in February and occasionally, in January, March or even November). This minimum is below 60–70 mm in areas above 800 m and below 40–50 mm in the low depression areas, representing up to 6 % of the annual precipitation total.

Figure 7.19 shows the regional distribution of monthly total precipitation across the Romanian Carpathians.

Seasonally, precipitation distribution is largely explained by the patterns of the synoptic moisture transport in interaction with the underlying topography of these mountains. This interaction reflects the regional distribution patterns of the annual

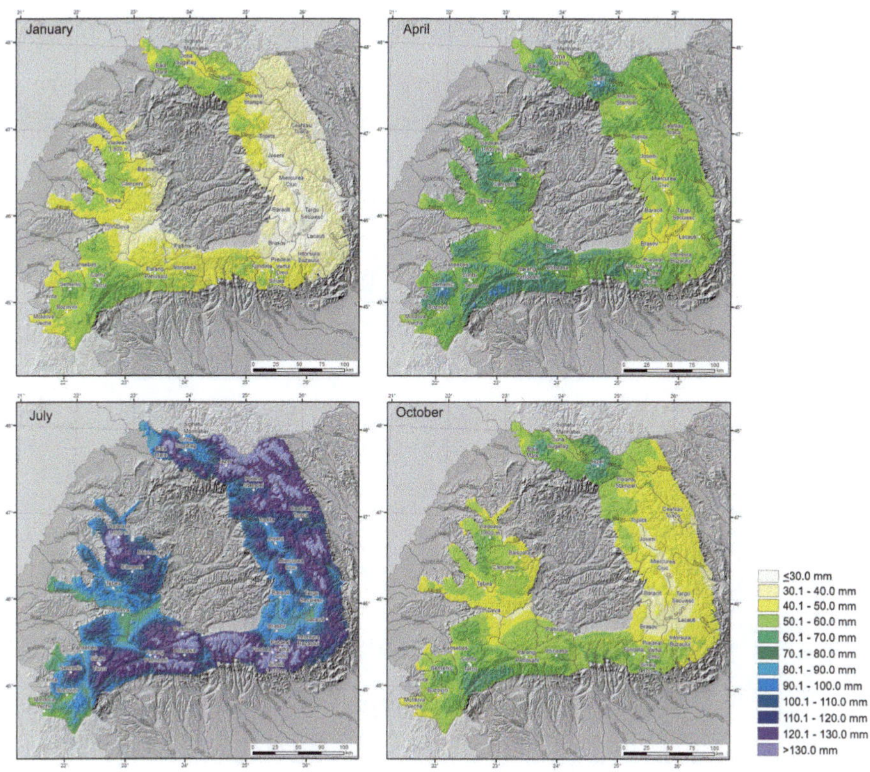


**Fig. 7.19** Monthly distribution of precipitation across the Romanian Carpathians

frequency of wet days[4] across the Romanian Carpathians: higher in the Southern and Western Carpathians (113–124 days) and lower in the Eastern ones (about 107 days). In response, the seasonal precipitation show fairly comparable amounts in the Southern and Western Carpathians, in connection with the influence of the westerly and southerly moist airflows, and considerably diminished values in the Eastern Carpathians.

Summer holds the greatest share in the annual totals (29–66 %). Summer amounts almost double the winter ones (the driest season). On average, the wettest season accumulates between 290 and 440 mm, at above 800 m and between 190 and 300 mm, in the areas below 800 m (Fig. 7.20). In this season, elevation explains 73 % of the vertical zonation of precipitation across the region. In high-elevation areas above 1,800 m, the summer totals stay below or around 400 mm, almost doubling the amounts specific to depression areas. The altitude effect on summer precipitation distribution is less visible than in winter (20 %), as significant amounts result frequently from rainfalls associated with deep convective thunderstorms.

[4] The threshold of 1.0 mm was chosen to define a wet day.

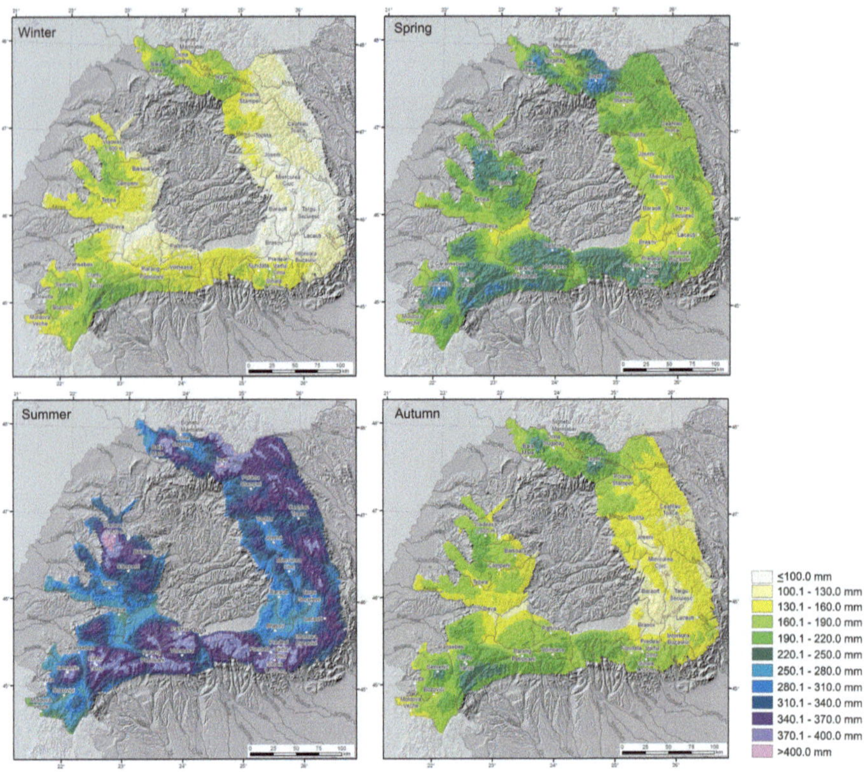


**Fig. 7.20** Seasonal distribution of precipitation across the Romanian Carpathians

Regionally, summer precipitation is the greatest in the Southern Carpathians (about 360 mm), a high-elevation cross-barrier for the warm and humid airflows from the Mediterranean Sea. These high-moisture content airflows are particularly responsible for the occurrence of orographic precipitating clouds and heavy convective rainfalls in this season. In the other Romanian Carpathian regions, summer precipitation range on average between 270 and 290 mm.

Under a prevailing anticyclonic regime, winter precipitation account only up to 35 % in the annual total. Precipitation falls are generally of frontal origin, peak accumulations being associated with cyclonic storms.

In the areas up to 1,400–1,500 m precipitation falls are rather shallow under persistent stratiform cloud cover and temperature inversion layer. Regionally averaged, winter precipitation amounts reach about 150 mm in the Southern and Western Carpathians and only 106 mm in the Eastern Carpathians (Fig. 7.20). In winter, elevation may explain 74–88 % of the precipitation spatial distribution across the Carpathians. Winter precipitation changes with elevation from about 63–150 mm, in depression areas to 95–235 mm, in areas above 800 m.

In spring the cyclonic activity gradually intensifies, particularly since March, being associated with frequent invasions of maritime air masses of polar (from west and north-west) or tropical origins (from south or south-west). The change in atmospheric flow patterns during this season results in a visible increase of spring precipitation (relative to winter ones) up to 160–320 mm, in areas above 800 m and 130–200 mm, in those below 800 m. Regionally, spring precipitation accumulates on average more than 200 mm only in the Southern and Western Carpathians, while in the Eastern ones, does not exceeding 170 mm.

Autumn is the second dry season of the year, due to the increased frequency of anticyclonic conditions and the weakening of thermal convection. This season accumulates on average about 180–190 mm in all the Southern and Western Carpathian and only 140 mm in the Eastern Carpathians (Fig. 7.20). The share of autumn amounts in the annual precipitation totals does not exceed 45 % at any of the selected weather stations of the region.

### *7.3.4 Precipitation Type*

The fraction of annual precipitation falling in solid or liquid forms is highly altitude dependent (Fig. 7.21). Table 7.10 summarizes these relationships at regional level across the Romanian Carpathians.

The share of the two main types of precipitation to the yearly number of precipitation days explains the zonation patterns of hydrological regimes across the Romanian Carpathians, as a function of the dominant feeding mechanisms of mountain rivers. According to the classification developed by Pardé (1955), the main types of hydrological regimes within the Romanian Carpathians are:

- *the pluvial regime*, where rainfalls prevails over the year, with high water in winter and low water in summer; this regime is typical for all the rivers of the areas below 800–1,000 m altitude;
- *the pluvio-nival regime*, where the liquid falls prevails (50–60 %) over the solid falls (40–50 %), at least 7 months per year, with a frequency of at least 10 days per month; these areas extend up to 1,500–1,700 m in the Southern and Eastern Carpathians and 1,400–1,500 m in the Western Carpathians.
- *the nivo-pluvial regime* where the solid falls are slightly dominant (50–60 %) over the liquid ones; this regime is generally characteristic of the areas of 1,800–2,000 m height;
- *the nival (subnival) regime*, where runoff is mostly controlled by snowfall and snowmelt. In these areas, the fraction of solid falls exceeds 60 %, particularly during the November-April interval, with a frequency of at least 10 days per month; this type of regime is a common feature of the alpine areas above 2,200–2,300 m.

The hydrological regime reflects the transitional climate of the region (between temperate and continental), with four distinct seasons, and with various climate influences – oceanic in the western part, Mediterranean in the south-west, Baltic in

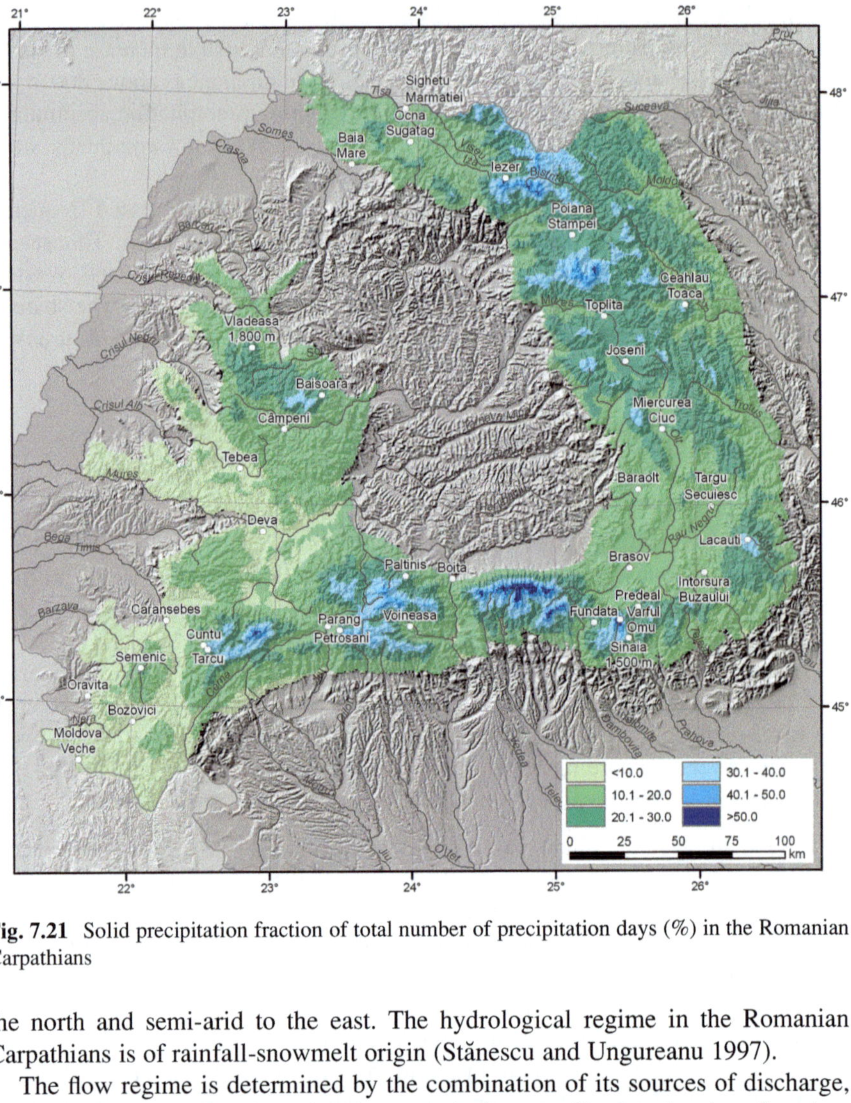


**Fig. 7.21** Solid precipitation fraction of total number of precipitation days (%) in the Romanian Carpathians

the north and semi-arid to the east. The hydrological regime in the Romanian Carpathians is of rainfall-snowmelt origin (Stănescu and Ungureanu 1997).

The flow regime is determined by the combination of its sources of discharge, which depends on climatic and anthropogenic factors affecting the river flow in a given catchment. The temporal evolution of the mean discharge presents important variations, usually determined by the characteristics of the climatic parameters that are generating the river flow, namely precipitation, air temperature, evapotranspiration and snow cover.

The intra-annual streamflow variability is directly linked to the pluviometric regime, and there are four characteristic periods that can be distinguished, corresponding to the seasons. The high mountain basins present 15 % of the total yearly discharge in winter, 40 % in spring, 30 % during summer and 15 % in autumn (Diaconu and Şerban 1994).

**Table 7.10** The relationship between the precipitation type fraction (%) and altitude in the Carpathians

| Romanian Carpathian branches | Altitude (H) – solid fall fraction (S) relationship | | | Altitude (H) – liquid fall fraction (S) relationship | | |
|---|---|---|---|---|---|---|
| | Gradient (% 100 $m^{-1}$) | Equation | Statistical significance | Gradient (% 100 $m^{-1}$) | Equation | Statistical significance |
| Eastern Carpathians | +2.0 | S = 3.66 + 0.02*H | $R^2 = 0.80$, $p < 0.01$ | −2.0 | L = 96.34–0.02*H | $R^2 = 0.80$, $p < 0.001$ |
| Southern Carpathians | +2.0 | S = −1.59 + 0.02*H | $R^2 = 0.88$, $p < 0.001$ | −2.0 | L = 101.6–0.02*H | $R^2 = 0.88$, $p < 0.001$ |
| Western Carpathians | +1.0 | S = 2.46 + 0.01*H | $R^2 = 0.85$, $p < 0.001$ | −1.0 | L = 97.54–0.02*H | $R^2 = 0.85$, $p < 0.001$ |

### *7.3.5 Precipitation Absolute Records*

The regional histograms of daily precipitation reveal a rather low probability of high precipitation accumulations (Fig. 7.22). The daily amounts of up to 10 mm are most likely to be registered across the region, regardless of elevation (93–94 %).

The regional distributions of daily precipitation are not 'heavy-tailed'. The extreme behaviour depicting the occurrence of heavy precipitation events (HPEs) is relative to values higher than at least 20 mm/day – a threshold set to assess the occurrence of very heavy precipitation days (WMO 2009). This threshold roughly corresponds to the 95th percentile value at more than 60 % of the selected weather stations. Overall, HPEs have a very low probability across the Carpathians (below 0.5 %) and they account for up to 24 % in the total number of wet days (above 1.0 mm). Comparative statistic of the regional frequency of HPEs (HPE20, R30 and R50mm) is given in Table 7.11.

Strong convective systems and intense summer cyclonic storms are most responsible for the occurrence of such values in the row-data, as precipitation tends to

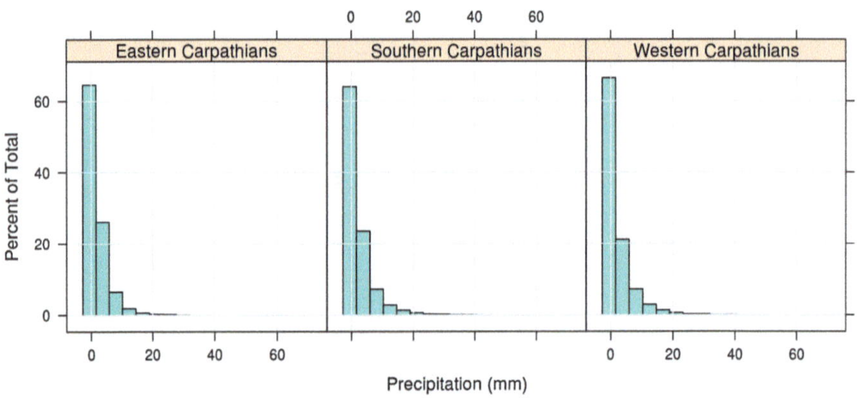


**Fig. 7.22** The regional histograms of daily precipitation in the Romanian Carpathians

**Table 7.11** Share of HPEs in the total number of wet days (%) across the Romanian Carpathians

| Regions | R20mm | R30mm | R50mm | Number of wet days |
|---|---|---|---|---|
| Eastern Carpathians | 12.7 | 6.0 | 3.3 | 106.9 |
| >800 m | 6.2 | 2.9 | 1.5 | 125.9 |
| <800 m | 4.3 | 1.3 | 0.2 | 98.5 |
| Southern Carpathians | 16.8 | 13.1 | 10.4 | 123.5 |
| >800 m | 13.3 | 9.2 | 7.0 | 128.7 |
| <800 | 6.9 | 2.7 | 0.4 | 102.7 |
| Western Carpathians | 23.9 | 17.5 | 13.8 | 112.6 |
| >800 m | 17.4 | 12.9 | 10.8 | 137.0 |
| <800 | 4.9 | 1.5 | 0.2 | 100.4 |

**Table 7.12** Absolute 1-day precipitation amounts (mm/date of occurrence) measured until 2010 in the Romanian Carpathians

| Months | Eastern Carpathians | Southern Carpathians | Western Carpathians |
|---|---|---|---|
| January | 59.9/January 17, 1961 (Lăcăuți) | 126.0/January 2, 1966 (Țarcu) | 38.8/January 8, 1963 (Caransebeș) |
| February | 43.8/February 22, 1980 (Iezer) | 81.5/February 10, 1984 (Voineasa) | 44.1/February 14, 1978 (Moldova Veche) |
| March | 50.6/March 5, 2006 (Iezer) | 96.9/March 23, 2007 (Predeal) | 48.8/March 31, 1982 (Semenic) |
| April | 71.7/April 1, 1927 (Ocna Șugatag) | 110.0/April 24, 1933 (Păltiniș) | 69.3/April 18, 1966 (Semenic) |
| May | 124.4/May 13, 1970 (Baia Mare) | 103.0/May 29, 1966 (Cuntu) | 77.5/May 11, 1973 (Semenic) |
| June | 105.2/June 18, 1949 (Joseni) | 136.0/June 12, 1965 (Cuntu) | 107.0/June 12, 1965 (Semenic) |
| July | 61.4/July 12, 2005 (Joseni) | 262.0/July 19, 1934 (Deva) | 122.0/July 30, 1971 (Semenic) |
| August | 93.7/August 25, 1977 (Lăcăuți) | 102.0/August 12, 1974 (Cuntu) | 122.6/August 12, 1974 (Semenic) |
| September | 89.4/September 17, 1957 (Lăcăuți) | 118.4/September 19, 2005 (Sinaia 1,500 m) | 127.0/September 1, 1941 (Caransebeș) |
| October | 100.0/October 22, 1990 (Ceahlău-Toaca) | 121.9/October 20, 1964 (Sinaia 1,500 m) | 66.6/October 10, 1980 (Semenic) |
| November | 74.6/November 10, 2010 (Iezer) | 76.7/November 7, 1961 (Voineasa) | 110.0/November 8, 1948 (Țebea) |
| December | 57.4/December 27, 1995 (Miercurea Ciuc) | 74.8/December 7, 1990 (Sinaia 1,500 m) | 53.8/December 12, 1995 (Vlădeasa 1,800 m) |

concentrate over short periods. The extreme behavior of HPEs is sampled by rare but very high daily precipitation amounts of more than 200 mm, as measured in the NMA network. The absolute 1-day precipitation amounts in the Romanian Carpathians totaled 204.2 mm/July 19, 1970 at Cuntu (1,456 m) in areas above 800 m and 262.0 mm/July 19, 1934 at Deva (240 m), in depressions (Table 7.12) and resulted from long-lasting precipitation events of high intensity, in connection with strong retrograde cyclonic activity.

The lower bound of the regional histograms of daily precipitation depicts the 'extreme dry domain' (Fig. 7.22). The frequency of dry days exceeds 60 % in all the three Carpathian regions, showing a great tendency to dry extremes all over the region. Regionally, the annual frequency of dry days (below 1.0 mm) is very high, ranging between 240 and 260 days across the three regions of the Romanian Carpathians. Generally, the areas below 800 m have the greatest share in the regional averages, where dry days may account for up to 250–280 days in the entire year. In the areas above 800 m, the annual frequency of dry days does not exceed 250 days.

## 7.4 Wind

Wind is the most dynamic element of mountain climate with discontinue action, highly fluctuating in time and space. The direction and strength of the synoptically driven winds across the Romanian Carpathians are widely controlled by the activity of the main high and low pressure systems operating at European scale: the Azores (all-year-round) and the East-European anticyclones (highly active during the winter months) and the Icelandic and the Mediterranean cyclones (all-year-round). Above timberline (at about 1,700–1,800 m), the wind direction and speed on site scale appear more closely connected to the atmospheric flow in the free troposphere, while below it the airflows are subject to the topographic effect and local surface roughness (e.g. vegetation).

### *7.4.1 Surface Wind Direction*

The high-elevation sites, with open meteorological platforms, located above timberline, were preferentially used to describe the annual and seasonal characteristics of surface winds in relation to some typical flow patterns, as the wind roses show a visible deflection of the upper level flow with height decrease, under the strong topographic effect.

The westerly wind direction prevails in all high-elevation areas above 1,700–1,800 m of the Romanian Carpathians, largely corresponding to the lower layers of zonal atmospheric flow (Fig. 7.23). In alpine areas above 2,500 m (Vf. Omu station), where airflow dynamics are taking place in conditions close the free troposphere, the westerly winds are dominant 8 months/year, generally from January to March (23–25 %), May to August (17–19 %) and in December (23 %), reaching an annual average frequency of about 21 %. At this particular height, the southwesterly winds are the secondary prevailing wind flows over the September-November interval (21–26 %) and in April (22 %), as controlled by the activity of low pressure Mediterranean systems with Trans-Balkan trajectories. In the areas of 1,700–1,900 m, which mark the transition between forest and subalpine vegetation belts in most of the Romanian Carpathians, the annual frequency of westerly winds may exceed 40 % only at the mountain sites directly exposed to the zonal atmospheric flow over the Apuseni Mountains (40.7 % at Vlădeasa 1,800 m, 1,836 m a.s.l.) and the Eastern Carpathians (45.8 % at Ceahlău-Toaca, 1,897 m a.s.l.). In these areas, the westerly flows are dominant all year (Fig. 7.24), reaching peak frequencies (more than 45 %) of 3–7 months during the cold half of the year: September-March (Ceahlău-Toaca) and December-February (Vlădeasa 1,800 m). The westerly winds prevail also in the Curvature sector of the Eastern Carpathians, favoring the occurrence of foehn effects on the outer side of these mountains. In this sector, the annual frequency of westerly winds is only 27 % (Lăcăuți, 1,776 m), even though their influence is dominant up to 9 months of the

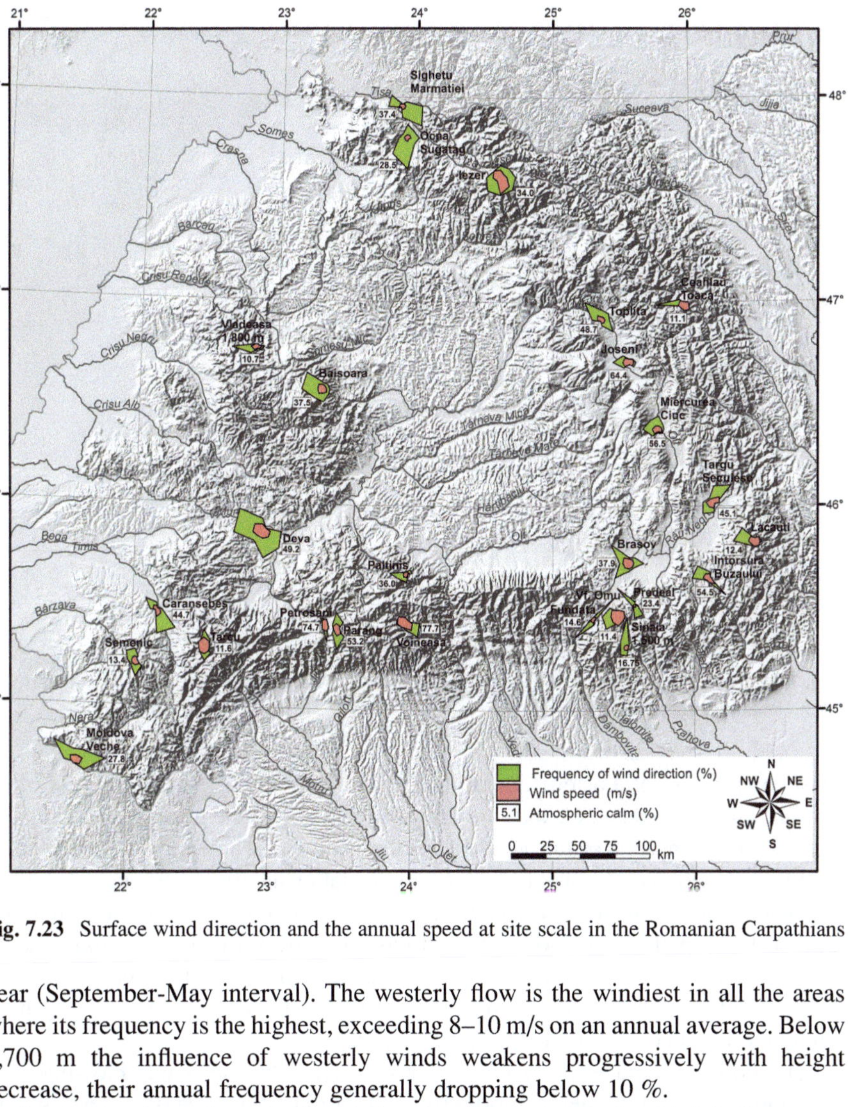


**Fig. 7.23** Surface wind direction and the annual speed at site scale in the Romanian Carpathians

year (September-May interval). The westerly flow is the windiest in all the areas where its frequency is the highest, exceeding 8–10 m/s on an annual average. Below 1,700 m the influence of westerly winds weakens progressively with height decrease, their annual frequency generally dropping below 10 %.

Southwesterly and southerly winds are often associated with Mediterranean frontal lows advancing towards eastern and north-eastern Europe on Trans-Balkan trajectories. The overall 'Mediterranean flow' reaches the peak annual frequencies (above 25 %) only in some high-elevation areas of the Southern Carpathians (Vf. Omu 26.8 %) and Banat Mountains (Semenic 29.1 %), particularly exposed to the southwesterly and southerly advection, but also in some mountain corridors with a general south-north orientation (e.g. Fundata 36 %, in the Rucăr-Bran Corridor). Seasonally, the Southern and Western Carpathians show the highest occurrence frequency of these winds, mostly in autumn (14–15 %), winter (12 %) and spring (12–13 %) and less in summer (10–11 %). The 'Mediterranean flow' is

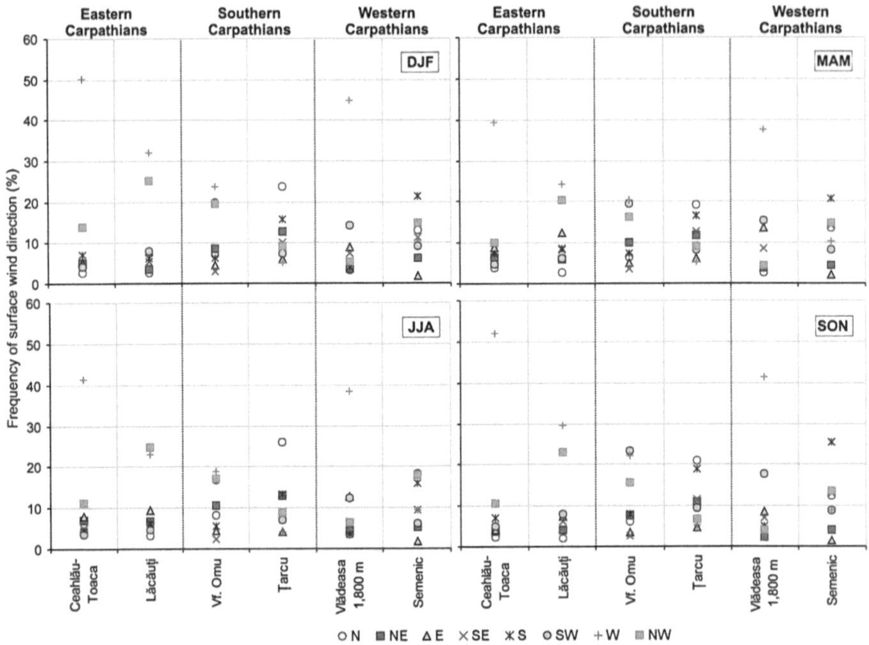


**Fig. 7.24** Seasonal frequency of wind direction at some sites with open meteorological platforms of the Romanian Carpathians

least common in the Eastern Carpathians (generally below 7 %), regardless of the season. The southeasterly winds are also associated with the passage of Mediterranean lows with normal or retrograde trajectories or with warm and dry (continental) summer advection of North African origin. The latter situations were more common in some summers of the last decade (e.g. 2007, 2012), resulting in persistent and intense heat spells in the southern and eastern extra-Carpathian regions and on the southern flank of the Eastern Carpathians in their Curvature sector (generally below 1,700 m). However, the frequency of the southeasterly winds is in general limited to less than 7 % in most Carpathian areas.

The northerly and easterly circulations depicted by the winds blowing from the northern (northwesterly, northerly and northeasterly) and eastern sectors (northeasterly, easterly and southeasterly) have a low annual frequency at most sites located above timberline, generally below 35 % and 20 %, respectively. Locally, in some high alpine areas of the Southern Carpathians above 2,000 m, some wind directions show notable enhancements relative to the prevailing westerly flow characteristic at this height e.g. at Ţarcu site located on the ridge line of the Ţarcu Mountains (at about 2,200 m), the northerly wind direction prevails most of the year, from December to September (above 18 %); at this particular station the annual frequency of northerly winds is 22 %, followed by the southeasterly and southerly directions (11–16 %), whereas the frequency of westerly airflow is significantly lower (5 %).

The topographic-induced flow can be considered characteristic only at local scale, giving no insight into the overall regional flow patterns. The terrain-forced flow along the main axes of mountain corridor, deep and narrow valleys or depression areas could generate significant deflections of the prevailing wind direction. Valley channeling with anti-winds explains the high annual frequency of some flow types: e.g. the southwesterly and northeasterly winds (above 30 %) at Fundata site, located in the Rucăr-Bran Corridor (1,384 m a.s.l.); the northwesterly, southerly and southeasterly winds (15–30 %) at Predeal site, in the Prahova Valley (1,090 m a.s.l.); the easterly and northwesterly winds (above 18 %) at Moldova Veche site, in the Danube Gorge (82 m a.s.l.).

### *7.4.2 Surface Wind Speed*

Mountain topography strongly influences the strength of the boundary layer flows as well as, the distribution patterns of wind velocity (Barry 2008; Mortensen and Petersen 1998; Etienne et al. 2010). Whiteman (2000) stressed that complex terrain exhibits major discontinuities in flow speed at local scale, due to the great contribution of surface roughness elements (e.g. peaks, trees, boulders), channeling effects or thermally-induced circulations.

Despite the likely disturbing effect on the surface wind flow of some topographic factors (e.g. altitude, slope, curvature, landform type), it is only elevation which is statistically significant (52 %) in explaining the vertical zonation of wind speed within the Romanian Carpathians. Wind speed (WS) increases with height, reaching peak values on mountain ridges, in conditions close to the free troposphere.

Regionally, the areas higher than 1,800–2,000 m of the Eastern and Southern Carpathians, exhibit the most dynamic climate (Fig. 7.25). In these areas the average winds are strong above 9–10 m/s on a multi-annual basis. Above timberline, the prevailing airflow directions are in most cases the windiest, exceeding in general 10 m/s: e.g. the westerly winds at Ceahlău-Toaca (10.7 m/s) and Vlădeasa 1,800 m sites (10.1 m/s); the northerly (11.8 m/s) and the northeasterly winds (10.3 m/s) at Ţarcu site; the northwesterly (10.5 m/s) and southerly winds (10.7 m/s) at Vf. Omu site.

Wind-induced stresses on local vegetation of windy areas is significant, giving rise to anatomical and morphological changes to isolated trees and in upper timberline trees. Holtmeier (2009) highlighted the disturbing role of winds on tree growth (both physiologically and mechanically) in mountain timberlines, which contribute to the development of tree deformations (e.g. asymmetry, slanting), providing evidence of the locally varying wind speeds and directions. Stoenescu (1951) emphasized that tree deformation is a valuable indicator of local boundary airflow dynamics, considering that the "flagged" trees act like true "biogeographical anemometers", particularly in areas with prevailing strong and persistent winds, blowing mostly from one direction (Fig. 7.26).

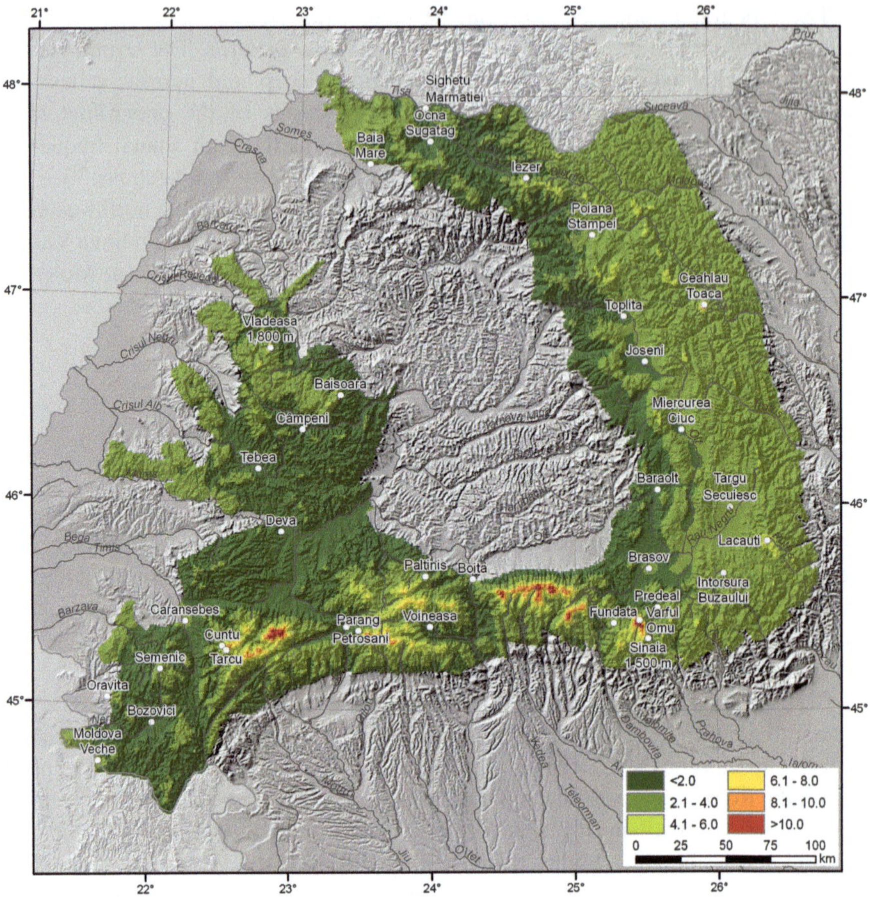


**Fig. 7.25** Annual average speed of surface winds (m/s) in the Romanian Carpathians

**Fig. 7.26** "Flagged" common beech trees/hill at south-west of Suseni – Harghita county, wind direction NE (*left*) and south-easterly "flagged" spruce trees in the upper timberline of the Siriu Mountains, the Curvature Carpathians (*right*) (Photo: C. Costandache – *left* and M. Micu – *right*, with permission)

The moderately windy areas, which appear partially exposed to the prevailing winds, are those located around 1,700–1,800 m elevation, at the transition between forest and subalpine vegetation associations. In these areas, the average WS ranges between 5 and 8 m/s (e.g. Lăcăuţi 7.6 m/s, Vlădeasa 1,800 m 6.7 m/s, Semenic 5.1 m/s). Below timberline, the local topographic features and surface roughness strongly modify the characteristics of the dominant airflow. From in situ measurements, these effects are highly visible below 1,500–1,600 m, where the average WS is generally lower than 3–4 m/s.

Wind speed increases significantly in narrow mountain areas (e.g. valleys, mountain corridors), as the streamlines tend to concentrate along their main axis. Ţâştea et al. (1990) and Bogdan (1994) investigated the frequency of strong winds in relation to pressure changes between the northern and southern extremities of the Olt River corridor, crossing the central part of the Southern Carpathians from north to south. The authors found pressure differences of up to 3–5 mb between the two extremities of the corridor, explaining the increasing frequency of strong winds above 24 m/s at Boiţa site (located in the northern extremity of the corridor), up 5 % in the November-March interval.

The daily surface wind speed regime across the Romanian Carpathians is dominated by light winds up to 2 m/s with occurrence probabilities of about 56 % in the Eastern Carpathians, 42–45 % in the Southern and Western Carpathians (Fig. 7.27). The extreme wind tail, depicting values higher than 10 m/s (the regional average of the 95th percentile level), has lower occurrence probabilities of 1.0–1.7 %, in the Eastern and Southern Carpathians and slightly higher ones, of about 3.0 %, in the Western Carpathians.

Seasonally, the changes in large-scale circulation result in major variations of WS at site scale. These changes are highly visible in areas above 1,700–1,800 m, particularly in winter and summer (Fig. 7.28). In winter, the airflow dynamics over the Carpathian mountain range is most significant, when the thermo-baric gradients are usually the steepest due to advective situations (January and February), resulting in frequent stormy winds. The high-elevation areas above 1,800 m of

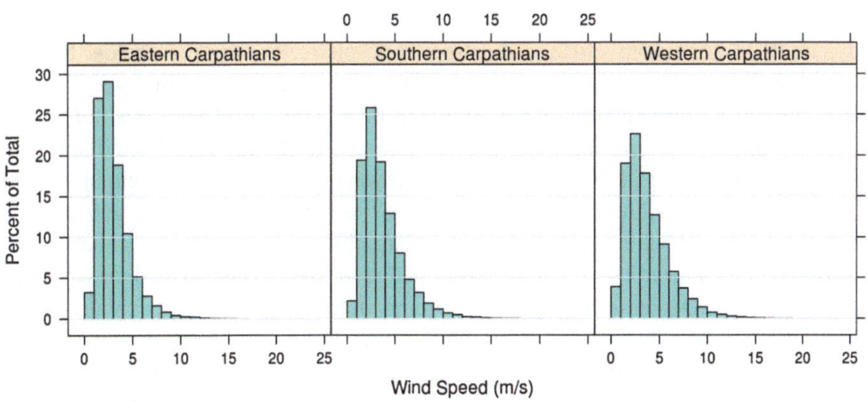


**Fig. 7.27** Comparative regional histograms of daily wind speeds in the Romanian Carpathians

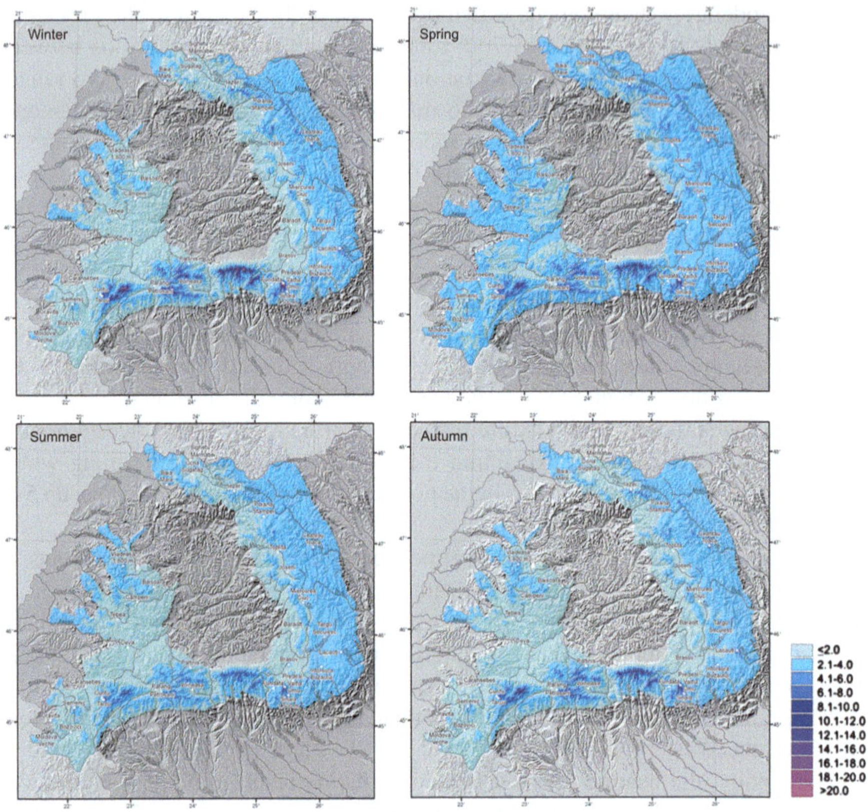


**Fig. 7.28** Distribution of seasonal wind speed (m/s) in the Romanian Carpathians

the Eastern and Southern Carpathians are particularly windy, from October to even April, the WS ranging from 8.0 to 13.9 m/s. Noteworthy, are the Ceahlău-Toaca (1,897 m – the Eastern Carpathians) and Vf. Omu sites (2,504 m – the Southern Carpathians), which are the windiest all year, as they best reflect the airflow dynamics in conditions close to the free troposphere. In the areas below 800 m, winter WS does not exceed 2 m/s, particularly in the intra-mountainous depression areas under the influence of stable stratification of cold air. The intensification of cyclonic activity at mid-latitudes is closely linked to the increased wind speed in spring, more visible in March and April. However, in spring, WS remains up to 2 m/s below the winter average. The May-September interval is in general, the calmest interval of the year, all over the Carpathians. The advective situations of warm and moist air transport, at the eastern edge of Mediterranean cyclones towards the Pannonian Depression, are very frequent during this interval, explaining the intensification of southerly winds at site scale in the Banat Mountains and on the southern flank of the Southern Carpathians, as well as the channeling effect along some transversal mountain corridors, with favorable orientation. Although the convective situations are highly frequent during the summer months, they are associated with low WS, as the surface pressure changes are not significant. On average, summer

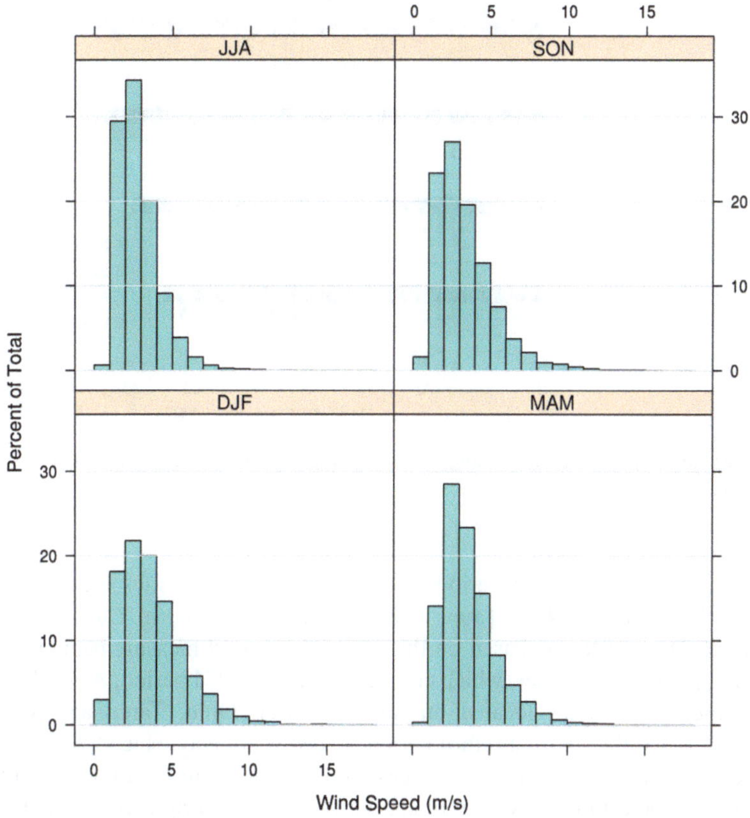


**Fig. 7.29** Seasonal frequency of daily wind speed (*WS*) across the Romanian Carpathians

winds rarely exceed 2 m/s in the areas below 800 m and 5–7 m/s in those above 800 m, except for the wind-sheltered ones, where WS stays below 3 m/s.

Figure 7.29 illustrates the seasonal frequency of daily WS across the Romanian Carpathians, showing the prevalence of light winds (below 5 m/s), particularly in summer (94 %). The extreme wind tail depicting WS values above 10 m/s, which roughly corresponds to the 95th percentile level at most mountain sites located above 1,400–1,500 m, is slightly visible in winter, spring and autumn, but its occurrence probability is limited to less than 1 %.

## *7.4.3 Maximum Wind Speed*

The Romanian specialist literature lacks regional studies aimed at providing accurate statistics of the frequency of wind gust speeds across the Romanian Carpathians. A valuable contribution to the knowledge of gusty winds at local scale in the Romanian Carpathians belong to Neacşa et al. (1973), who investigated

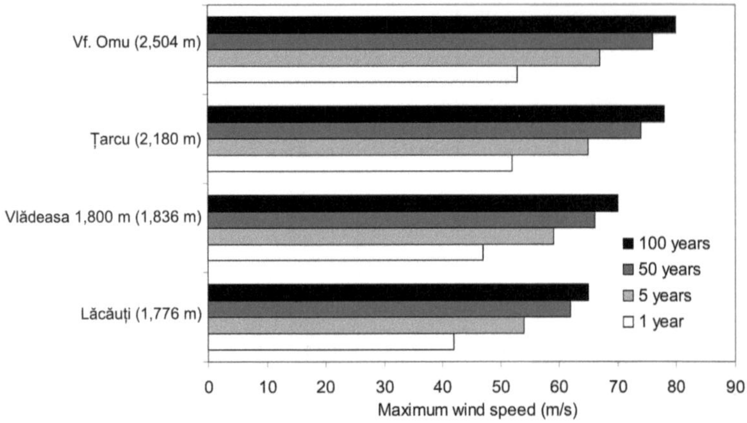


**Fig. 7.30** Return periods corresponding to wind gust speeds at some high-elevation sites of the Romanian Carpathians

the $WS_{max}$ and their return periods for five high-elevation sites located in open terrains (Vf. Omu 2,504 m; Ţarcu 2,180 m; Vlădeasa 1,800 m 1,836 m; Lăcăuţi 1,776 m). This work was based on the daily wind speed measurements using the swinging plate anemometer, which covered the 1962–1974 period. Accordingly, at the high-elevation sites located on ridgelines (Vf. Omu, Ţarcu and Vlădeasa 1,800 m), the $WS_{max}$ with a return period of 100 years may exceed 70 m/s, while for the other site (Lăcăuţi), these values were estimated to be around 53–65 m/s (Fig. 7.30). The estimated wind gust speeds were found fairly comparable to those in the High Tatra Mountains at above 2,000 m. Later, Ţâştea and Neacşa (1974) determined the return periods of maximum wind speed for the 600–2,500 m elevation range within the Bâlea Lake-Capra area (the Făgăraş Mountains – the Southern Carpathians), using 2-min wind speed measurements over the 1961–1969 period. The general findings of the study suggest a gradually increase with height of the estimated $WS_{max}$ with a return period of 100 years, from 40 m/s at about 1,400–1,500 m to more than 70 m/s above 2,000 m (Fig. 7.31).

Investigating the relationships between pressure gradients and gusty winds frequency in different regions of Romania (including the Romanian Carpathians), Ţâştea and Lorentz (1985) showed that the northerly, southerly, westerly and easterly circulations are most responsible for the occurrence of extreme wind episodes.

The spatial distribution of wind gust speeds across the Romanian Carpathians is shown in Fig. 7.32. The highest $WS_{max}$ values (hereby considered as the maximum value of the four daily observations) exceeded 40 m/s (Beaufort scale 12) in all windward ridge areas of the Romanian Carpathians located above 1,800 m, e.g., Ceahlău-Toaca, Vf. Omu, Vlădeasa 1800 (Table 7.13). However, Stoenescu (1951)

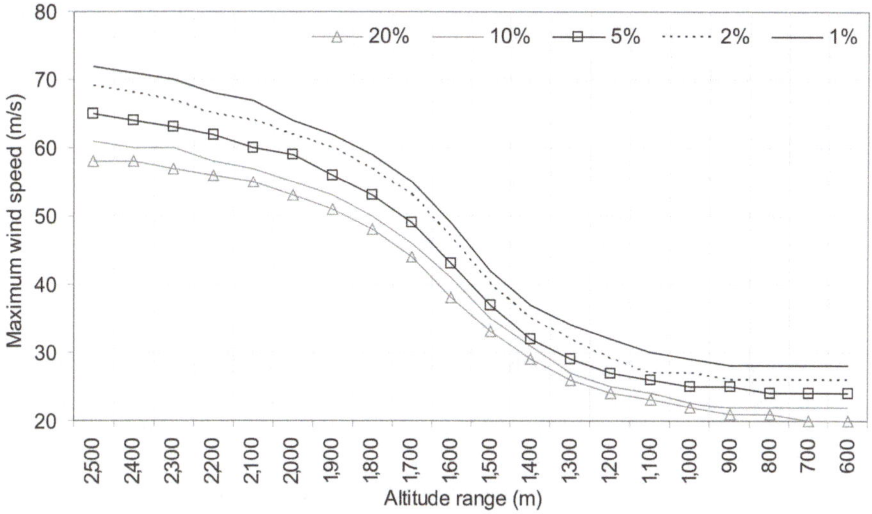


**Fig. 7.31** Wind gusts with different return periods in the Bâlea Lake-Capra area (the Făgăraş Mountains – the Southern Carpathians)

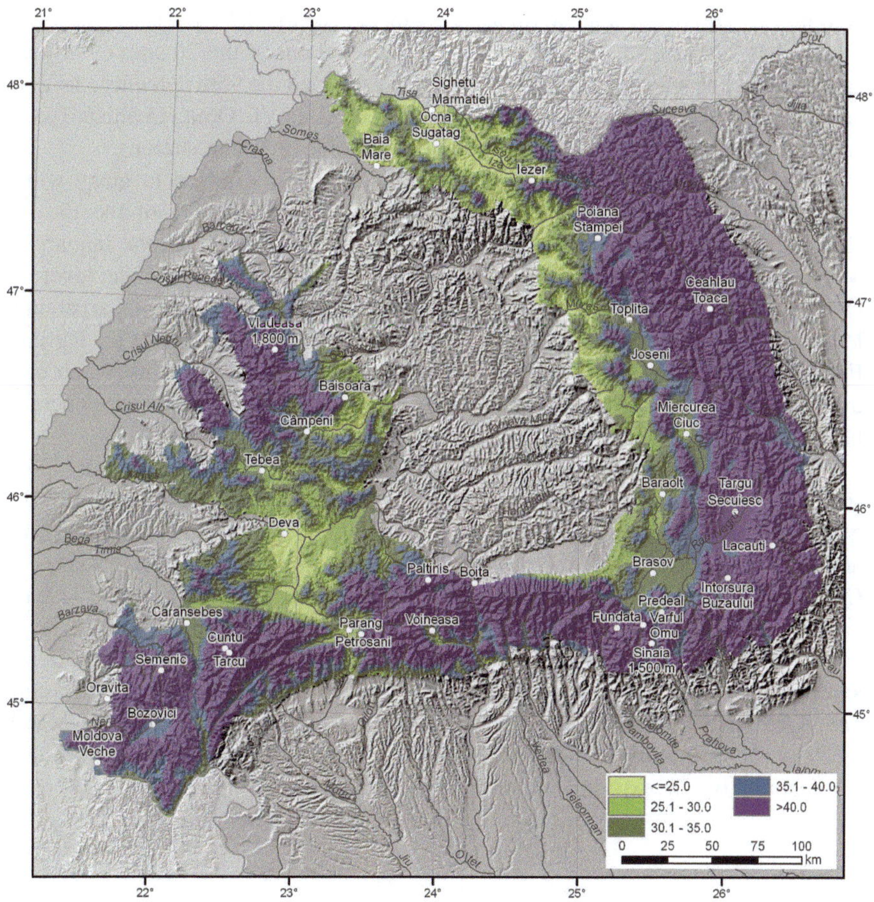


**Fig. 7.32** Maximum wind speed distribution in the Romanian Carpathians (m/s)

**Table 7.13** Regional $WS_{max}$ record (m/s) in the Romanian Carpathians (1961–2010)

| Romanian Carpathians regions | Areas above 800 m | Areas below 800 m |
|---|---|---|
| Eastern Carpathians | >40 m/s (Ceahlău-Toaca, 1,897 m) | 34 m/s (Braşov, 534 m) |
| Southern Carpathians | >40 m/s (Vf. Omu, 2,504 m) | 26.4 m/s (Petroşani 607 m) |
| Western Carpathians | >40 m/s Vlădeasa 1800, 1,836 m) | 38.6 m/s (Moldova Veche, 82 m) |

reported the occurrence of an outstanding violent gusty wind episode in the alpine area of the Southern Carpathians (in the Bucegi Mountains). Accordingly, during the night of November 27–28, 1932, wind speed exceeded 80 m/s at Vf. Omu station, a level well above Beaufort scale 12.

The lowest $WS_{max}$ are specific to wind-sheltered sites located in intra-mountainous depression areas, ranging in general between 25 and 35 m/s. The channeling and shooting effects observed in some mountain/valley corridors and open depression areas could explain the extreme behavior of surface wind speed, exceeding 30 m/s (Beaufort scale 11) at some low-elevation sites below 600 m (e.g. 38.6 m/s at Moldova Veche, 82 m a.s.l.; 34 m/s at Braşov, 534 m a.s.l.; 31.5 m/s at Caransebeş 240 m a.s.l.). At most weather sites (20 out of the 26 sites), the wind gust speed records over the 1961–2010 period exceeded the 95th percentile level of daily wind speed and exceptionally, the 99th level at Vf. Omu, Ceahlău-Toaca, Lăcăuţi, Ţarcu, Parâng, Vlădeasa 1,800 m and Moldova Veche stations.

On a monthly basis, the mountain areas particularly exposed to gusty winds exceeding 40 m/s (Beaufort scale 12) are those above 1,800 m of the Eastern Carpathians (from October to May) and Western Carpathians (from January to March), as well as those higher than 2,000 m of the Southern Carpathians (over the October-April interval). Exceptionally, in some areas of the Southern Carpathians located below timberline at about 1,300–1,500 m (e.g. Sinaia 1,500 m, Păltiniş, Fundata), the wind gusts decreased below 20 m/s during the November-February interval. Below 800 m, the extreme winds recorded fairly lower speed, generally below 15 m/s all year round.

The monthly frequency of strong winds, when WS exceeds 16 m/s (Beaufort scale 7) provides evidence of the seasonal interval prone to extreme winds and gusty episodes. Extreme winds become more likely with height, reaching peak annual frequencies in windward areas above 1,700 m: more than 50 days at Ceahlău-Toaca and Vf. Omu sites and 25–35 days at Ţarcu, Lăcăuţi, Vlădeasa 1,800 m sites. In the low-elevation areas, such winds are very rare (up to 2 days/year). Seasonally, the frequency of strong winds across the Romanian Carpathians shows visible differences at regional scale. The high-elevation areas of the Eastern and Southern Carpathians are subject to strong winds from October to March (at least 5 days/month), with peak frequency in winter (21–27 days/season). Comparatively, the exposure to strong winds of the areas higher than 1,800 m of the Western Carpathians is limited to the November-March interval, when their frequency is up to 2–4 days/month. Generally, the May-September interval is least favorable to strong wind occurrences all over the Romanian Carpathians (up to 3 days/month).

Wind speed higher than 16 m/s is a critical value, causing extensive damage to the forests of the Romanian Carpathians (Bogdan and Coşconea 2010). The analysis of decennial frequency of strong winds over the 1961–2010 period revealed that the 1960s and 1970s experienced the greatest frequency of extreme wind cases, which were related to several intense cyclonic episodes (e.g. 1961–1963, 1965 and 1970). The gusty wind episodes during the 1960s and 1970s gave rise to some of the greatest windthrow losses, particularly in the forests of the Eastern Carpathians (e.g. Marcu et al. 1969; Popa 2000, 2005; Macovei 2009; Bogdan and Coşconea 2010). Analyzing the wind regime in Romania, (Vespremeanu-Stroe et al. 2012) found a statistically significant shift in storm activity at country scale at the end of the 1970s and beginning of the 1980s. A strong link between storminess and the North Atlantic Oscillation (NAO) index with a dominant positive phase over the 1979–2003 interval, was also emphasized within this study.

### 7.4.4 Atmospheric Calm

The daily wind speed distribution (Fig. 7.27) shows that the probability of calm events is generally below 4 % all over the Romanian Carpathians: 2.1 % in the Southern Carpathians, 3.2 % in the Eastern Carpathians and 3.9 % in the Western Carpathians.

The atmospheric calm (wind speed lower than 1 m/s) reaches peak frequencies of 50–70 %, in the closed intra-Carpathian depression areas (Bogdan 1985; Bogdan and Dragotă 1997). In open depression areas and transverse mountain corridors, the local airflows subject to channeling effect explain the lower values of calm situations, generally below 45 % (e.g. Moldova Veche 27.8 %, Braşov 37.9 %, Caransebeş 44.7 %) (Fig. 7.23). Above timberline, the atmospheric calm does not exceed 14 % (e.g. Ţarcu 11.6 %, Vf. Omu 11.4 %, Ceahlău-Toaca 11.1 % and Vlădeasa 1,800 m 10.7 %). The local topography and vegetation-induced roughness (forest vegetation) could induce local increases of calm frequency in some mountain areas particularly where wind-sheltered (e.g. Iezer 34.0 %, Păltiniş 36 %).

Peak probabilities of calm winds are characteristic of summertime (particularly in August), which may go generally up to 20 % in the high-elevation areas above 1,700 m and up to 65 % in the wind-sheltered low-elevation areas. In winter the calm frequency is significantly reduced (mostly in December and February), below 10 % at the high-elevation sites subject to the prevailing winds and increased above 50–60 % in the depression areas.

### *7.4.5 Local Winds*

Temperature contrasts within mountain areas are particularly important in deriving diurnal thermally-induced winds. A number of previous studies investigated the diurnal local flows in different areas of the Romanian Carpathians (e.g. Bâzâc 1981, 1983; Bogdan 1994; Stăncescu 1983; Clima 2008). Bâzâc (1983) emphasized that the greater the thermal contrasts along valley axes, the higher the frequency and speed of diurnal winds (up-valley and down-valley winds, also known as 'valley breezes' and 'mountain breezes'), yet without exceeding 5 m/s. The diurnal circulation manifests decoupling from the prevailing airflow, reaching peak speeds over the 01.00–07.00 a.m. interval at most mountain sites. These types of local flow are most likely to develop in clear-sky and calm conditions (under the influence of stable air masses), particularly in spring, summer and autumn. Diurnal winds are particularly important for the timing of snowmelt in lower mountain valley sectors, by producing earlier thawing. Bâzâc (1981, 1983) indicated that the thickness of the air layer subject to diurnal flows varies between 500–1,000 m for 'valley breezes' and 200–300 m for the 'mountain' ones.

On the leeside of some mountains, oriented perpendicularly to the prevailing flow directions, the warm and dry downslope winds (the *foehn*) appear as a common pattern, favoring the decrease of local temperature, air humidity and cloudiness, associated with a low precipitation probability and earlier snowmelt. The mechanisms of foehn winds and the climatic effects of these local winds have been investigated in detail by several authors (e.g. Stăncescu et al. 1986; Bogdan and Niculescu 1990; Bogdan 1993; Ion-Bordei 2008). Stăncescu (1983) pointed out that the occurrence of foehn winds are strongly linked to the moisture content of the prevailing air masses, the speed of airflows over the mountain ranges and to the height of the orographic barrier.

Figure 7.33 illustrates foehn-prone areas across the Romanian Carpathians according to Bogdan (1993). The occurrence of foehn winds favors the presence of several thermophilic vegetation formations (e.g. lilac, smoke tree, south European flowering ash) in the local flora of foehn-prone areas (Muică 1989).

The Curvature Region, including both the Carpathian the Subcarpathian sectors, is distinguished as one of the most representative in Romania for foehn occurrences. Bogdan (1993) asserted that in this region, foehn is likely all-year-round, reaching peak frequencies in winter, spring and summer, under the influence of prevailing westerly flows. Studying the Carpathian topography effect on atmospheric flow, Bordei-Ion (2008) demonstrated that the concave form of the Eastern Carpathians in the Curvature sector (the funnel-like shape) favors westerly flow to cross over the upwind slopes, enhancing precipitation on the windward side and foehn effects on the leeward. Stăncescu (1983) outlined that, in this particular region, the foehn produces visible warmer and drier weather, with sharp temperature rises during the day (up to 10–15 °C), clear sky, few precipitation and gusty flows (of up to 30 m/s).

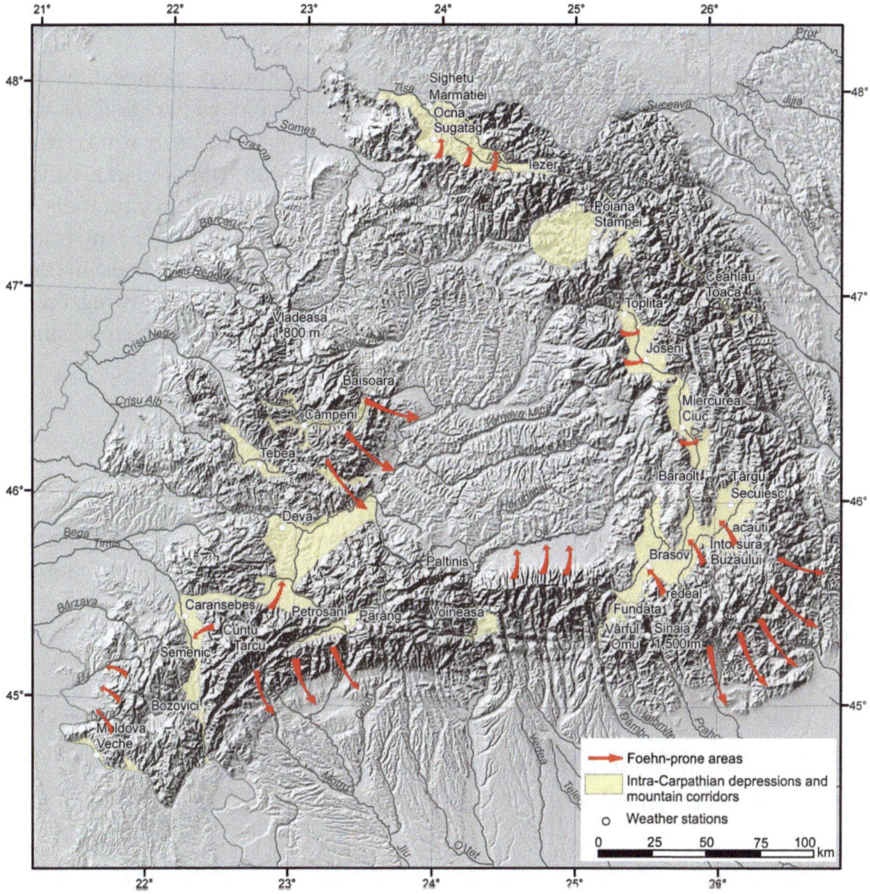


**Fig. 7.33** Foehn-prone areas across the Romanian Carpathians (Adapted after Bogdan 1993, with permission)

A typical foehn wind episode, observed on the lee side of the Curvature Carpathians and Subcarpathians that developed under the influence of northwesterly advection, was investigated by Bogdan and Niculescu (1990). Its occurrence was recorded over consecutive nights during the first 10 days of December 1988, with peak intensity on the night of December 6–7. This foehn episode was particularly intense from 18.00 p.m. to 02.00 a.m. and its characteristics have been well reflected by the meteorological measurements provided by a local weather station (Pătârlagele, 286 m), namely: northerly wind speeds of 10 m/s, temperature rise of about 3 °C, relative humidity below 50 % and a snowpack (about 25 cm) melting off. The effects of this foehn episode were felt down to Făurei station (located southeastward in the Romanian Plain, at 54 m), where the air temperature increased by 1 °C. Tilinca et al. (1976) has previously shown that such advection usually triggers strong foehn winds on the lee side of the Apuseni Mountains

(Western Carpathians) and lighter, dry and warm winds on the lee sides of Eastern Carpathians (in the Curvature sector) and Southern Carpathians.

Other characteristic flow situations, inducing significant changes in local weather types are also: the *Nemira*, a branch of the cold winter easterly wind affecting the eastern extra-Carpathian regions (the so-called *Crivăț* wind), which crosses the Eastern Carpathians through low-elevation (transversal) valleys, reaching the Brașov Depression; the *Munteanul*, which brings heavy rainfalls and hailstorms under the influence of Mediterranean cyclones which are retrograded, blowing from the Buzău Mountains (the Eastern Carpathians) towards the northern and central Bărăgan Plain (the Romanian Plain); the *Pietrarul*, blowing on the northwestern slopes of the Piatra Craiului Mountains (the Southern Carpathians).

## 7.5 Snowfall and Snowpack

Snowpack is the quantitative expression of winter precipitation usually dominant in solid form beyond 1,700 m altitude in the Romanian Carpathians. Amstrong and Brun (2008) emphasized that snowfalls are not a property of snowpack, but exert a key role in its initiation and accumulation on the ground.

### *7.5.1 Atmospheric Circulation Patterns Related to Snowfall Events*

Generally, the atmospheric circulation conditions most frequently related to snowfall and heavy snowfall occurrences in the Romanian Carpathians are: Atlantic or Mediterranean moisture transport in interaction with cold air advections from high latitudes of eastern continental Europe origins, large pressure gradients, resulting in strong winds and uplift due to the Carpathian orography and 500 hPa configurations (e.g. cut-offs). According to Gugiuman and Stoian (1972), who provided the first comprehensive study on snowfall and snow cover regimes in the Romanian Carpathians, cold and wet advection originating from the North-Atlantic and sub-polar regions are atmospheric patterns commonly related to snowfall and heavy snowfall occurrences over the October–April interval, particularly in the Eastern Carpathians. In addition to this, Bogdan (1978) asserted that the north-easterly flows are responsible for the earliest snowfalls in the northern mountain regions of Romania.

Several works investigated the synoptic and meteorological conditions relevant in explaining the occurrence of snow load in the forests of the Romanian Carpathians, as result of heavy snowfalls (e.g. Haring and Iuga 1970; Barbu 1979, 1982; Ichim and Barbu 1981). Their findings highlight the importance of the southerly (Mediterranean) airflows encountering the Carpathian barrier, of the interactions between the westerly and easterly circulations on the southern and eastern flanks of the Southern and Eastern Carpathians, as well as of air temperature oscillations around the freezing point (0 °C) and light winds or calm conditions during snowfall deposition.

Micu and Mărăşoiu (2009) investigated the heavy snowfall climatology and the synoptic conditions related to their manifestations in the Southern Carpathians at above 1,000 m. In this study, a heavy snowfall day was considered to occur when snow depth increased by at least 20 cm relative to the previous day. The snow amount of 20 cm was related to the threshold depicting the 'yellow code' warning level of the National Meteorological Administration (NMA), for 24 h fresh-snow accumulations in areas above 800 m. The study showed that moist airflows and the troposphere vertical temperature profile depicting variations from 0 °C at site's level to −10 °C on cloud peaks (Altostratus, Nimbostratus, Stratocumulus, Stratus, Cumulus and Cumulonimbus) are highly important factors favouring the heavy snowfall occurrence in these mountains. The southward expansion of the Icelandic Low and the airflows moving eastward from the south and south-west of the Mediterranean basin were found to enable a great moisture transport over the Carpathians, particularly in the October-April interval. Furthermore, the presence of some cold nuclei at 500 hPa geopotential height proved to play a key role in ensuring abundant snowfalls, particularly in May and September.

### *7.5.2 Snowfall Occurrence Interval*

The occurrence probability of the first snowfall in autumn ($SF_{autumn}$) and last snowfall in spring ($SF_{spring}$) changes also, considerably with height. In autumn, the first snowfalls across the Romanian Carpathians appear 1–2 months later (37–54 days) in the foothill areas, below 800 m, than in the areas higher than 800 m. On average, the first snowfalls occur from early September, in high-elevation areas above 2,000 m, to early-to-mid November, in the depression areas (Table 7.14). Regionally, the snowfalls appear the earliest (before September 1st) in the northern parts of the Eastern Carpathians and in the alpine areas of the Southern Carpathians (Fig. 7.34). However, significant deviations relative to the average occurrence range of $SF_{autumn}$ have been recorded all over the Romanian Carpathians, generally covering the August to December interval. The earliest autumn snowfalls were recorded in early August in mountain areas above 1,700 m (Vf. Omu, Ţarcu, Vlădeasa 1,800 m, Iezer) and in late August and early September, in those below 800 m (Întorsura Buzăului, Petroşani, Caransebeş). Such early occurrences were recorded in some of the snowiest or coldest years of the study period (e.g. 1963, 1966, 1972–1973, 1975, 1996). The latest autumn snowfall was reported in December in the areas above 1,400–1,500 m of the Southern and Western Carpathians (Parâng and Semenic) and at about 900 m in the Eastern Carpathians (Poiana Stampei). The first snowfalls were delayed the most, until January or even March in some foothill areas below 600 m: e.g. January 1, 2005 (Sighetul

**Table 7.14** Average and extreme occurrence dates of snowfalls in the Romanian Carpathians

| Regions | Average dates | | Extreme dates (earliest/latest) | |
|---|---|---|---|---|
| | $SF_{autumn}$ | $SF_{spring}$ | $SF_{autumn}$ | $SF_{spring}$ |
| Eastern Carpathians | | | | |
| >800 m | Sep. 19–Oct. 28 | Apr. 24–Jun.19 | Aug. 1st, 1966 (Iezer)/Dec. 16, 2000 (Poiana Stampei) | Mar. 18, 2010 (Poiana Stampei)/Jul. 29, 1966 (Iezer) |
| <800 m | Oct. 30–Nov. 12 | Apr. 11–May. 1 | Aug. 30, 1996 (Întorsura Buzăului)/Jan. 1st, 2005 (Sighetul Marmației) | Jan. 25, 2002 (Toplița)/May 14, 1980 (Joseni, Toplița, Miercurea Ciuc, Târgu Secuiesc, Ocna Șugatag) |
| Southern Carpathians | | | | |
| >800 m | Aug. 17–Oct. 14 | Apr. 30–Jul. 11 | Aug. 1st, 1973, 1975 (Vf. Omu, Țarcu) and 1977 (Țarcu)/Dec. 11, 2000 (Parâng) | Jan. 28, 2002 (Cuntu)/Jul. 31, 2008 (Vf. Omu) |
| <800 m | Nov. 12–15 | Apr. 12–May. 1 | Oct. 4, 1972 (Petroșani, Voineasa)/Jan 21, 1980 (Voineasa) | Jan. 1st, 2002 (Voineasa)/May 9, 1987 (Petroșani, Voineasa) |
| Western Carpathians | | | | |
| >800 m | Sep. 4–Oct. 10 | May 14–Jun. 23 | Aug. 1st, 1963 (Vlădeasa 1,800 m)/Dec. 7, 1961 (Semenic) | Apr. 1st, 2007 (Semenic)/Jul. 31, 1979 (Vlădeasa 1,800 m) |
| <800 m | Nov. 12–26 | Apr. 30–Jun. 21 | Oct. 1st, 1963 (Caransebeș)/Apr. 20, 1969 (Deva) | Jan. 21, 2002 (Moldova Veche)/May 9, 1987 (Câmpeni) |

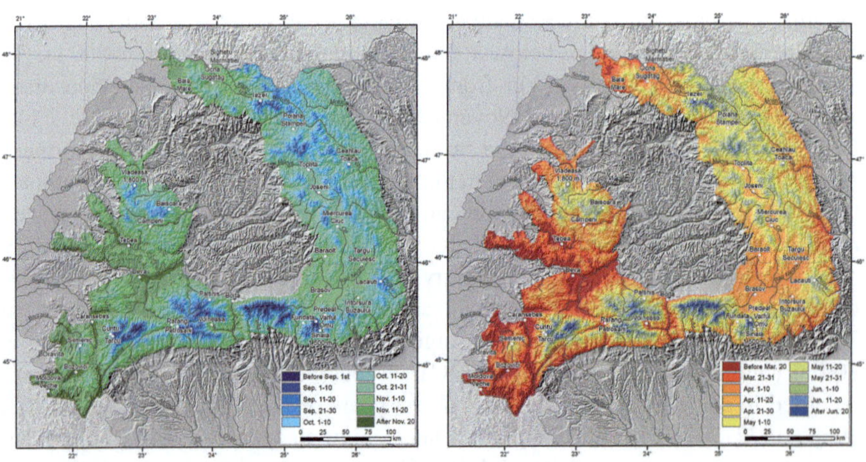

**Fig. 7.34** Average calendaristic dates of $SF_{autumn}$ (*left*) and $SF_{spring}$ (*right*) in the Romanian Carpathians

Marmaţiei, 275 m a.s.l.); January 21, 1980 (Voineasa, 573 m a.s.l.); March 1st, 1980 (Deva, 240 m a.s.l.). Relative to its average occurrence date, the latest $SF_{autumn}$ across the Romanian Carpathians recorded delays of up to 2½ months (40–70 days) in the areas above 800 m and up to 3½ months (26–100 days), in those below 800 m. Most of these delays, were recorded in the last decade of the study period, particularly in the year 2000 (at 15 out of the 32 weather stations), one of the warmest years on record in these mountains. In alpine areas, the delays in the $SF_{autumn}$ occurrence ranged between 50 and 70 days, as the latest dates were recorded on October 6, 1994 (Vf. Omu, 2,504 m a.s.l.) and November 13, 1994 (Ţarcu, 2,180 m a.s.l.).

In spring, the last snowfalls across the Romanian Carpathians are likely from late April to early July, in the areas above 800 m and from late March to mid-April, in depression areas (Table 7.14 and Fig. 7.34). Generally, the last snowfalls in the areas below 800 m could disappears two up to 1 month earlier than in those higher than 800 m, particularly in the Western Carpathians where winters are usually the mildest. The last snowfalls are likely until early July only in high-elevation areas of the Southern Carpathians (above 2,000 m) and from late April until mid-June in the rest of the areas above 800 m across the Romanian Carpathians. In summer, the snowfalls are usually light and produce mostly shallow snow accumulations. Important deviations from the last snowfall average date have been observed with elevation change. Noteworthy, is the fact that the last snowfalls disappeared the earliest in January, in some areas below 800 m (Topliţa, Voineasa, Moldova Veche stations) and exceptionally, at about 1,500 m in the Southern Carpathians (Cuntu station). A grouped signal indicating earlier $SF_{spring}$ is coming from the last decade of the study period, when 22 out of the 32 surveyed stations recorded the earliest spring snowfall particularly in the warm years of 2007 (6 sites), 2002 (5 sites) and 2010 (4 sites). Exceptionally, in 2002, the deviations from last snowfall average date exceeded 3 months in some areas of the Southern and Western Carpathians: 109 days on January 28, 2002 (Cuntu, 1,456 m a.s.l), 120 days on January 1, 2002 (Voineasa, 573 m a.s.l.), 151 days on January 21, 2002 (Moldova Veche, 82 m a.s. l.). At the end of the snow season, snowfalls are mostly light and may fall even if the snowpack melted, producing rather shallow and patchy snow accumulations. The latest $SF_{spring}$ recorded delays from the norm of up to 3 months, particularly in some years of the coldest and snowiest decades: 91 days in the Eastern Carpathians (Poiana Stampei, 923 m a.s.l.)/July 24, 1964; 71 days in the Southern Carpathians (Parâng, 1,548 m a.s.l.)/July 27, 1963; 64 days in the Western Carpathians (Semenic, 1,432 m a.s.l.)/July 18, 1991.

## *7.5.3 Snowfall Frequency*

The eastward transition from an Atlantic and Mediterranean climate, with milder winters (higher temperatures and more liquid precipitation) towards a more continental one, with colder winters and more and longer snowfall intervals, is generally

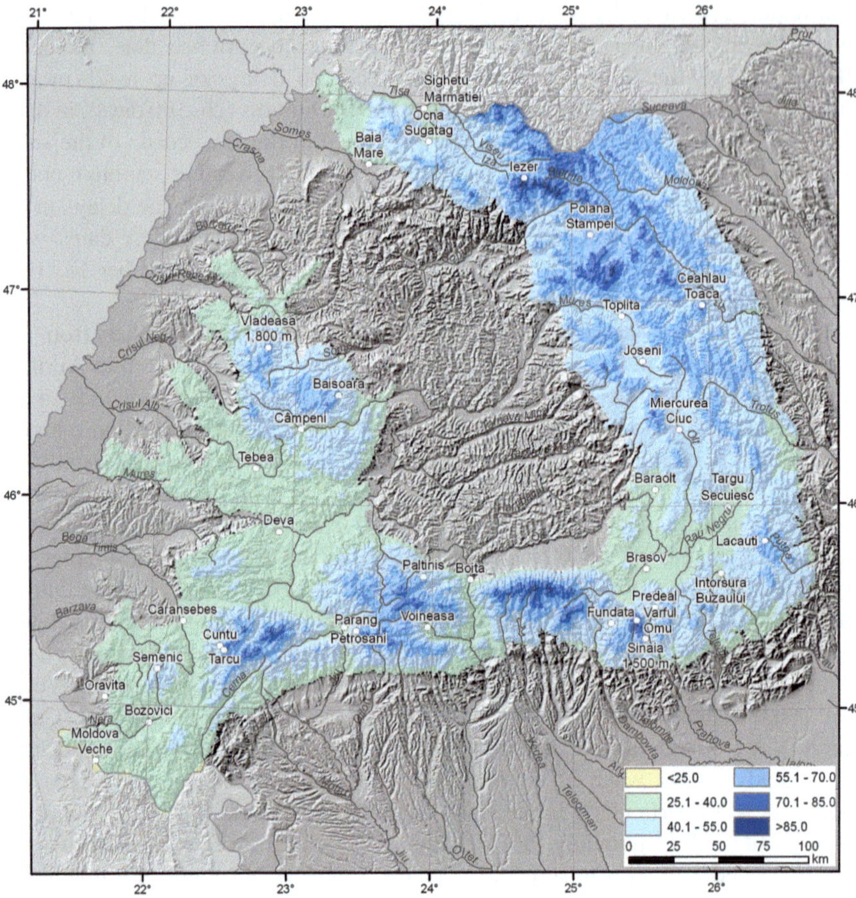


**Fig. 7.35** Spatial distribution of snowfall frequency (days) during the November–April interval

reflected by the spatial distribution of the snowfalls frequency during the extended cold season (November–April interval) across the Romanian Carpathians (Fig. 7.35). The frequency of snowfalls across the Romanian Carpathians is largely dependent on elevation, which produces a linear seasonal increase (November–April) at a rate of 0.47 days/100 m in the Eastern Carpathians and a quite similar one of 0.40–0.42 days/100 m in the Southern and Western Carpathians. Statistically, this influence is strong in all the three geographic divisions of the study region, elevation explaining 64–85 % of the snowfall spatial distribution. The strong increase in snowfall frequency with height is more evident at above 800 m, where such events have at least a double occurrence probability than in the foothill areas. This increase is generally related to the rain-snow transition.

The snowfall frequency increases from 24–46 days in the areas below 800 m to 56–96 days in those above 800 m. Regionally, the areas subject to frequent snowfalls of at least 60 days/year are those higher than 900 m in the Eastern

Carpathians, 1,000–1,100 m in the Southern Carpathians and above 1,300 m in the Western Carpathians.

Over the year, the contribution of the calendar winter months (December–February) to the total number of snowfall days is significant (at least 60 %), especially in the areas below 800 m. During the November–April interval, this contribution reaches 97–100 %. In this interval, the areas above 1,700–1,800 m in the Eastern (northern part) and Southern Carpathians appear as areas subject to frequent snowfalls, of at least 70 days/season. These areas are the coldest in the Romanian Carpathians and are open towards the direction of the main precipitation systems originating from northern and southern Europe. On the contrary, the snowfall frequency during this season is lower in the Western Carpathians, even at similar elevation levels, due to the prevailing south-westerly and westerly flows, which produce more liquid than solid precipitation. Peak monthly frequencies are specific to January in most areas and exceptionally, to December, only in those areas above 800 m of the Western Carpathians, where the share of snowfall events is about a half of the month duration (12.8–14.1 days/month).

### *7.5.4 Snow Showers*

The shower character of snowfalls is limited to less than 40 days/year all over the Romanian Carpathians. This phenomenon is very frequent during the November–April interval, accounting for 91 to even 100 % in the total number of cases throughout the year. Statistically, elevation controls only 42 % of the seasonal variation of snow showers (November–April) in the Southern Carpathians and up to 71 % in the Eastern and Western Carpathians, giving fairly comparable gradients of about 1 day/100 m.

The snow shower maximum is in March (at 22 out of the 30 surveyed stations) and less spatially spread, in April, at some sites located at over 1,400 m in the Southern Carpathians, at above 1,700 m in the Eastern Carpathians and 1,800 m in the Western Carpathians. This maximum is linked with the intensification of the cyclonic activity (normal and retrograde) in the Mediterranean and Black Sea basins. During the annual maximum, snow shower frequency does not exceed 10 days/month.

Regionally, the areas appearing highly exposed to snow showers (more than 20 days/November–April interval) are the alpine belt of some mountain massifs of the Southern Carpathians (e.g. the Făgăraş, Cindrel, Ţarcu Mountains), the western flanks of the Eastern Carpathians and the high-elevation areas of the Apuseni Mountains. In these areas, the local orography induces air mass uplift favouring strong vertical motions, which locally, result in abundant snowfalls in wintertime, usually associated with strong winds (Fig. 7.36). The low elevation areas (below 800 m) have in general the lowest frequency of snow shower events, as the surrounding mountains are shading them from large heavy snowfall events.

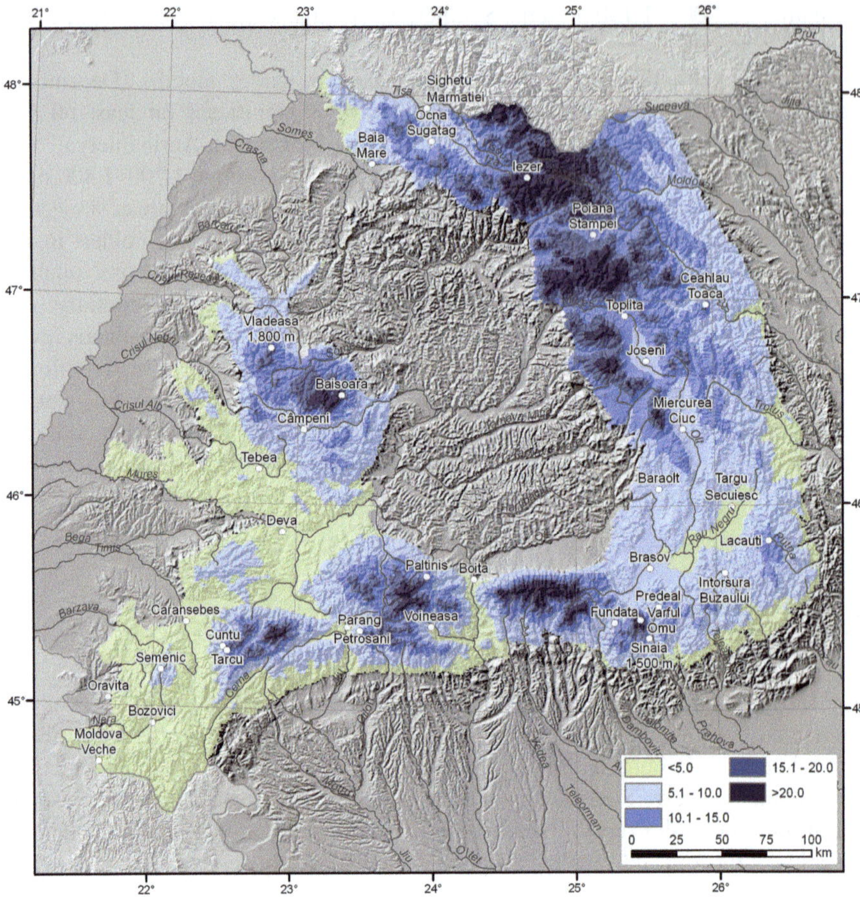


**Fig. 7.36** Frequency of snow showers (days) during the extended winter season (November–April)

Several works carried out in different mountain regions have investigated the climatology of heavy snowfalls, mostly in relation to the snow avalanche failure (e.g. Kocin and Uccellini 1990; Wild et al. 1996; Mote et al. 1997; Ancey and Charlier 1998; Esteban et al. 2005; Avalanche handbook 2006). These investigations suggest that fresh snow amounts of 10–30 cm and particularly those higher than 30 cm, accumulated in 24 h or shorter time-spans, are in general highly significant for snow avalanche releases. Romanian literature lacks studies on snow showers causing large snow accumulations, as fresh snow measurements started only in 2004, within the National Nivo-Meteorological Programme of the National Meteorological Administration at five sites located above 1,000 m in three massifs of the Southern Carpathians (the Bucegi, Postăvaru and Făgăraş Mountains). The overall aim of this programme is to monitor and forecast snow cover and avalanche risk in the surveyed Carpathian areas.

Despite their low occurrence probability, snow showers have the potential to trigger significant disruption of socio-economic activities, when the fresh snow accumulations are significant, being related to isolation of high-elevation villages, transportation disruption, damage to life-line infrastructure, limitation of access to ski resorts and their operation. The fresh snow accumulations over time-spans of 24 h are subject to three warning levels of the NMA. For areas higher than 800 m, the thresholds related to fresh snow amounts depicting these warning levels are: 20–40 cm ('yellow code'), 40–80 cm ('orange code') and over 80 cm ('red code').

Snow showers proved highly significant for large snow avalanches failures as shown by some works e.g. (Moţoiu 2008; Voiculescu 2009; Voiculescu et al. 2010). However, the specialist literature still lacks in studies aimed to deepen the knowledge of the relationship between fresh snow accumulations and snow avalanche failures within the Romanian Carpathians.

Snow showers could also entail great losses in forest-related activities and large disturbances in forest ecosystems, when the new snow amounts could exceed 40 mm/24 h or 50–60 mm/48 h (Barbu 1998). In general, coniferous forests proved to be particularly affected by snow load and some in situ tree characteristics proved to ensure the amplification of their vulnerability (e.g. slightly tapering stems, asymmetric crowns, rigid horizontal branching) (Haring and Iuga 1970; Barbu 1982). Barbu (2004) showed that the highest losses were encountered during the 1970s (1977, 1979), 1980s (1985) and 1990s (1993, 1995, 1996), when the most impacted forest ecosystems were the coniferous ones and inside them, mainly the trees lying outside their natural habitats. A comprehensive study on snow-induced forest damages was carried out at a national scale by Barbu (1998). The author made a hierarchy of the main geomorphic regions of Romania in respect to damage risk associated with snow load in the Romanian forests, using a probabilistic approach, considering the proportion of weather stations and rain gauges where the precipitation amounts was higher than 40 mm/24 h and 50–60 mm/48 h. The analysis was conducted for the January–April and October–December intervals and it was based on daily precipitation measured over the 1896–1955 period. Accordingly, three out of four exposure classes were assigned to different mountain areas within the Romanian Carpathians territory as follows: the 'very high' exposure class (above 66 %) to the Curvature Carpathians (Eastern Carpathians); the 'high' class (46–65 %) to the Southern Carpathians and the Banat Mountains (Western Carpathians); and the 'moderate' class (26–45 %) to the Maramureş and Bucovina Carpathians (Eastern Carpathians) and the Apuseni Mountains (Western Carpathins).

### 7.5.5 Snow Cover Interval

Favourable meteorological conditions for snow onset on the ground (e.g. persistent freezing, sufficiently abundant snowfalls, light winds), are usually encountered with a certain spatial pattern. The average snow onset dates vary from late

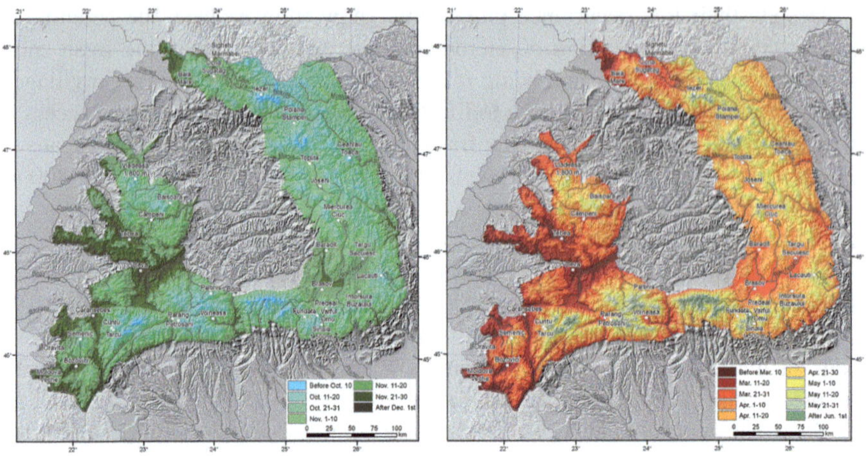


**Fig. 7.37** Average calendaristic dates of $SC_{autumn}$ (*left*) and $SC_{spring}$ (*right*) in the Romanian Carpathians

September in the areas higher than 800 m of the Eastern Carpathians to early December in foothill areas of the Western Carpathians (Fig. 7.37). On average, snowpack set on the ground a week up to 1 month later than the first average snowfalls.

The snowy and thermal character of some winters entailed significant deviations from the average snow onset date. The earliest $SC_{autumn}$ were recorded in high-elevation areas above 1,800 m, in early and late August (Table 7.15). In the rest of mountain areas, it occurred from early September to late October, generally 20–71 days in advance of the norm. Such deviations were characteristic of some snow-abundant and cool months of the study period e.g. August 1983, September 1996, October 1970, 1971, 1972. Autumn snowpack formed the latest after mid-November, at most sites located above 800 m and after mid-December, at those located in foothill areas. Exceptionally, the latest snow onsets are specific to several low-elevation areas (below 300 m) of the Western Carpathians, where the snowpack delayed it onset 58–94 days, until late February or even mid-February (e.g. January 30, 1998 – Moldova Veche 82 m; February 3, 21, 23, 2001 – Ţebea 273 m, Caransebeş 241 m and Deva 240 m). It is noteworthy that a great share of the latest snow onsets in the total number of observations belong to the 2000/2001 winter, which was one of the warmest winters within the region, as about 53 % of snow onsets have been recorded from November 2000 to February 2001 (with a high contribution of December 2000), in most areas of the Romanian Carpathians, regardless of the elevation. During this winter, the $SC_{autumn}$ delays vary between 42 and 94 days and were recorded particularly at the weather sites located in the Southern and Western Carpathians.

The snow cover season ends on average between mid-April and mid-June, in areas above 800 m and between late February and early April, in depression and low mountain areas (Fig. 7.37 and Table 7.15). However, the snowmelt is retarded

**Table 7.15** Average and extreme snow onset and snowmelting dates in the Romanian Carpathians

| | Average dates | | Extreme dates (earliest/latest) | |
|---|---|---|---|---|
| Regions | $SC_{autumn}$ | $SC_{spring}$ | $SC_{autumn}$ | $SC_{spring}$ |
| Eastern Carpathians | | | | |
| >800 m | Sep. 29–Nov. 3 | Apr. 17–May 23 | Aug. 30, 1998 (Ceahlău-Toaca)/Dec. 17, 2000 (Poiana Stampei) | Mar. 23, 2010 (Poiana Stampei)/Apr. 23, 1986 (Iezer) |
| <800 m | Nov. 10–Nov. 24 | Mar. 12–Apr. 8 | Oct. 4, 1972 (Joseni, Miercurea Ciuc)/Feb. 1st, 2001 (Joseni) | Jan. 14, 1972 (Braşov)/May 8, 1989 (Joseni, Baraolt) |
| Southern Carpathians | | | | |
| >800 m | Oct. 15–27 | Apr. 24–Jun. 22 | Aug. 2, 1989 (Ţarcu)/Dec. 17, 2000 (Sinaia 1,500 m, Păltiniş) | Mar. 26, 2007 (Predeal)/Jul. 29, 1979 (Vf. Omu) |
| <800 m | Nov. 20–21 | Mar. 20–23 | Oct. 4, 1972 (Petroşani, Voineasa)/Jan. 2, 2001 (Voineasa) | Jan. 29, 2002 (Voineasa)/Apr. 21, 1991 (Voineasa) |
| Western Carpathians | | | | |
| >800 m | Oct. 1–30 | Apr. 25–May 21 | Aug. 29, 1981 (Vlădeasa 1,800 m)/Dec. 12, 2000 (Băişoara, Semenic) | Mar. 28, 1994 (Băişoara)/Jul. 4, 1964 (Vlădeasa 1,800 m) |
| <800 m | Nov. 23–Dec. 8 | Feb. 24–Mar. 22 | Oct. 15, 2009 (Câmpeni)/Feb. 23, 2001 (Deva) | Jan. 20, 1989, 2008 (Moldova Veche)/Apr. 28, 1984 (Ţebea, Bozovici) |

considerably in high-elevation areas above 1,700–1,800 m, until mid-to-late May, and exceptionally, until mid June, at about 2,500 m in the Southern Carpathians (Vf. Omu). On average, snowmelting ($SC_{spring}$) occurs 6 up to 27 days before the last spring snowfall, in the areas above 800 m of the Eastern and Southern Carpathians and 13–33 days, in those of the Western Carpathians. In foothill areas, snowmelting is significantly earlier than the last snowfalls, the deviations varying from 8 to 50 days in the Eastern Carpathians to 20–42 days in the Southern Carpathians and to only 10–16 days in the mild Western Carpathians. In general, the snow cover season ends one and a half (in the Eastern and Southern Carpathians) to 2 months earlier (in the Western Carpathians) in foothill areas compared to those higher than 800 m.

The timing of snowmelt is related to the thermal character of some winter and spring months, in response to recent warming. In spring, mountain areas up to 1,400–1,500 m and most foothill areas of the Romanian Carpathians (particularly those of the Western Carpathians) are usually prone to sudden warming (high spring temperatures, frequently associated to rainfalls) and subsequent snow melting, as result of the frequent or persistent warm air advection of tropical origins (Mediterranean or North African), moving towards the Carpathian mountains range on a southerly or south-westerly flow component. Earliest onsets of snow melting in the Romanian Carpathians were recorded in some of the warmest spring and winter months of the study period, without stretching across large mountain areas (only up to three sites simultaneously): e.g. January 2002, February and March 2007, April 1986. Above 800 m, snow melting started the earliest since March (March 23–31), only at some sites located at 1,000–1,500 m (e.g. Poiana Stampei 923 m a.s.l., Sinaia 1,500 m, Cuntu, Băişoara). In intra-depression areas, the earliest onsets of snowmelting were observed from January 10 to February 28. Regionally, snowpack melted the earliest in the Western Carpathians, on January 10, 1989 and 2008 (Moldova Veche, 82 m a.s.l.). Generally, the earliest onsets of snowmelting were observed 29 up to 71 days before the average $SC_{spring}$.

Snow cover lasted the longest, until July, at only few high-elevation sites of the Romanian Carpathians: July 29, 1979 (Vf. Omu, 2,504 m a.s.l.); July 23, 1980 (Ţarcu, 2,180 m a.s.l.); July 10, 1998 (Iezer, 1,785 m a.s.l.); July 4, 1964 (Vlădeasa 1,800 m, 1,836 m a.s.l.). In the rest of the areas, the latest snowmelt records covered the late May to late June interval at above 800 m and the early April to mid-May, below 800 m. Regionally wide, the delays of the latest snowmelting relative to the norm, ranged between 21 and 53 days.

Altitude explains significantly (70–82 %) the zonation of the duration of the continuous snow cover interval (SCD) across the Romanian Carpathians, showing different magnitudes of the regional increases of snow cover duration as follows: 8 days/100 m in the Eastern and Southern Carpathians and 5 days/100 m in the Western Carpathians. The monthly statistics show that the snowiest interval of the year, when the snow cover days account for more than 80 % of the annual number of such days, is November–April, in areas above 800 m and December–February, in those below 800 m. In these intervals, snow is covering the ground at least half of the month, generally from 10 to 31 days.

Annually averaged, the continuous snow cover lasts the longest (at least 150 days/year) in the subalpine areas of the Eastern Carpathians (151 days – Ceahlău-Toaca, 1,897 m a.s.l.) and alpine ones of the Southern Carpathians (164 days – Vf. Omu, 2,504 m a.s.l.; 155 days – Ţarcu, 2,180 m a.s.l.). Comparatively, in the high-elevation areas of the Apuseni Mountains (Vlădeasa 1,800 m, 1,836 m a.s.l.) and exceptionally, at about 1,500 m in the Banat Mountains, the snow cover duration is only slightly more than 120 days/year. The length of continuous snow cover interval does not drop in general below 100 days in the areas above 800 m. In foothill areas, SCD stays below 70 days in the Eastern and Southern Carpathians, and well below this level (11–39 days), in those of the Western Carpathians (Fig. 7.38, Table 7.16).

The snowpack could be maintained continuously on the ground up to 114–140 days longer in the areas above 800 m than in the foothills. Regionally wide, the Western Carpathians have the shortest snow cover durations, while the Eastern Carpathians exhibit slightly longer snow cover durations than in the Southern Carpathians, highlighting the key role of atmospheric control in determining the very cold and persistent winter weather conditions, in both areas above and below 800 m.

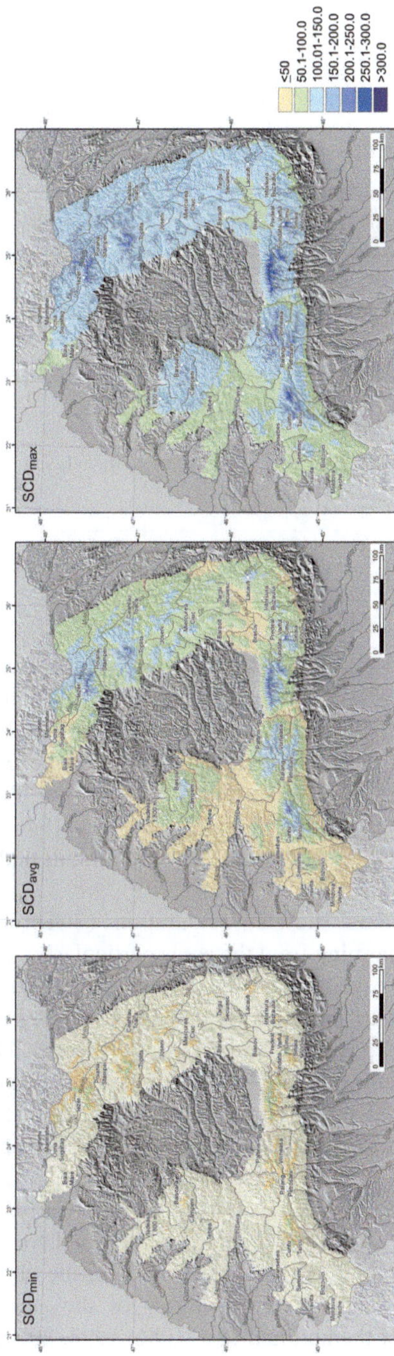


**Fig. 7.38** Snow cover duration (days) in the Romanian Carpathians

**Table 7.16** Average and extreme durations of continuous snow cover (days) in the Romanian Carpathians

| Romanian Carpathians regions | $SCD_{min}$ (the longest/the shortest) | $SCD_{avg}$ | $SCD_{max}$ (the longest/ the shortest) |
|---|---|---|---|
| Eastern Carpathians | | | |
| >800 m | 50 (Ceahlău-Toaca)/26 (Poiana Stampei) | 94.6–150.8 | 194 (Ceahlău-Toaca)/ 139 (Poiana Stampei) |
| <800 m | 11 (Toplița)/7 (Ocna Şugatag, Sighetul Marmației) | 36.8–76.8 | 131 (Miercurea Ciuc)/ 84 (Baraolt) |
| Southern Carpathians | | | |
| >800 m | 91 (Vf. Omu)/5 (Păltiniş) | 106.8–163.8 | 216 (Vf. Omu)/148 (Fundata) |
| <800 m | 9 (Voineasa)/4 (Petroşani) | 25.1–65.2 | 115 (Voineasa)/79 (Petroşani) |
| Western Carpathians | | | |
| >800 m | 35 (Vlădeasa 1,800 m)/13 (Băişoara) | 96.7–127.8 | 169 (Vlădeasa 1,800 m)/159 (Băişoara) |
| <800 m | 2 (Câmpeni, Caransebeş)/1 (Ţebea, Bozovici, Deva, Moldova Veche) | 11.5–39.3 | 111 (Câmpeni)/35 (Moldova Veche) |

The range of snow cover duration extended significantly over the 50-year period of observations, showing large deviations relative to the climatological norm (Table 7.13). The longest maximum snow cover durations ($SCD_{max}$), exceeding 190 days, are specific to the highest-elevation sites of the Eastern and Southern Carpathians e.g. 194 days at Ceahlău-Toaca, 216 days at Vf. Omu, 190 days at Ţarcu. Regionally, $SCD_{max}$ ranges between 84 and 194 days in the Eastern Carpathians, 79 and 216 days in the Southern Carpathians and 35–169 days in the Western Carpathians (Fig. 7.38).

On the contrary, minimum snow cover durations (SCDmin) provide clear evidence of recent climate warming from the areas particularly subject to winter warm spells, rapid snowmelt, frequent liquid falls in winter or local foehn effects. These areas corresponds mostly to some intra-mountainous depressions of the Southern and Western Carpathians, generally below 600 m, where SCDmin do not exceeds 4 days/year: 4 days (Petroşani 607 m a.s.l.); 2 days (Câmpeni 591 m a.s.l., Caransebeş 241 m a.s.l.); 1 day (Ţebea 273 m a.s.l., Bozovici 256 m a.s.l., Deva 240 m a.s.l. and Moldova Veche 82 m a.s.l.) (Fig. 7.38). In the areas above 800 m, SCDmin varies from 26–50 days in the Eastern Carpathians, to 5–91 days in the Southern Carpathians and to 13–35 days in the Western Carpathians. In these areas, an exceptionally short SCDmin was recorded in the Southern Carpathians (Păltiniş 1,453 m a.s.l.), reaching only 5 days. Nevertheless, the shortest SCDmin were recorded in several outstanding warm winters of the study period, when the snowmelt occurred earlier than normal (e.g. 1986, 1994, 2000, 2003).

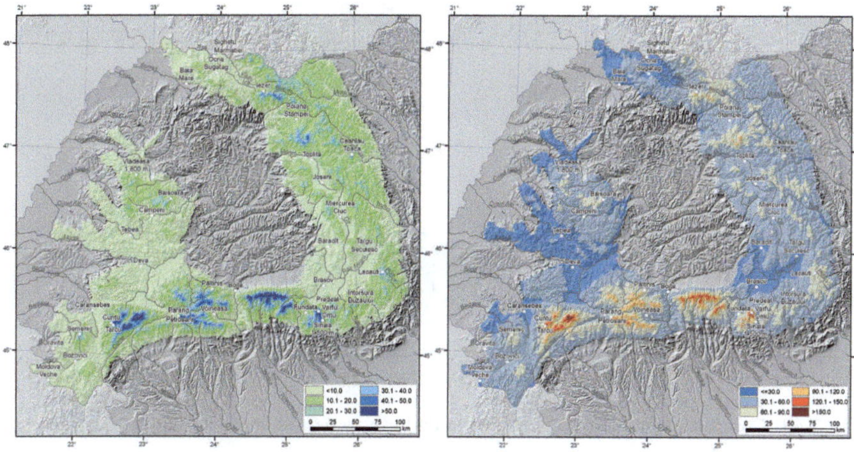


**Fig. 7.39** Spatial distribution of $HS_{avg}$ (*left*) and $HS_{max}$ (*right*) during the November–April interval in the Romanian Carpathians (cm)

## *7.5.6 Snow Depth*

The interactions between snowfalls and wind with the underlying terrain features and vegetation cover in mountain environments induce a great spatial heterogeneity in the patterns of snow accumulation on the ground. Wind redistribution effect dominates the snow depth spatial variability at 1–1,000 m scales, in dependence on both orography (terrain) and surface roughness (vegetation) (e.g. Elder et al. 1991; Winstral et al. 2002). Cheval et al. (2010) found that the monthly precipitation is more underestimated at high-elevations due to strong winds, particularly over the November-April interval, when solid falls prevail. Despite the significant redistribution effect of strong winds during the cold season, the areas above 1,800 m exhibit the largest snow accumulations across the Romanian Carpathians (Fig. 7.39).

The increase of cyclonic activity during the late winter to early spring interval, when snowfall events are frequently abundant, explains the occurrence of snow depth peak in this interval. At most sites (17 out of the 30 sites), generally located in the areas below 800 m of the Eastern Carpathians and at above 1,000–1,300 m in the Southern and Western Carpathians, the greatest average snow depths ($HS_{avg}$) over the year is recorded in February, in the range of 7–88 cm. For other Carpathian areas, including the high-elevation areas of the Eastern Carpathians (above 1,700 m) and one alpine site of the Southern Carpathians (Ţarcu), peak $HS_{avg}$ are specific to March (39–71 cm). Exceptionally, the greatest $HS_{avg}$ are characteristic of January (3–14 cm) only in some low-elevation areas of the Southern and Western Carpathians, generally below 600 and 300 m, respectively. Generally, average snow depths of at least 10 cm are observed from December to April or even May, in most areas above 800 m of the Eastern and Southern Carpathians, from December to March in the areas above 1,300 m of the Western Carpathians and in January and February in some intra-mountainous areas of 600–800 m of the

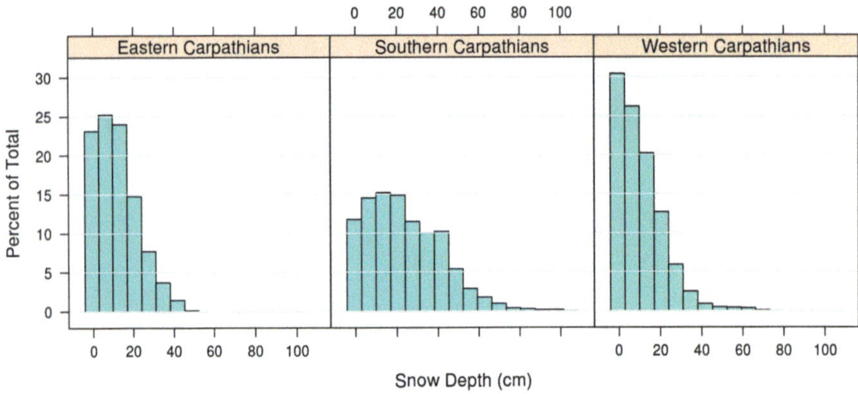


**Fig. 7.40** Comparative regional histograms of daily snow depths in the Romanian Carpathians

**Table 7.17** Highest maximum snow depth records ($SC_{max}$) in the Romanian Carpathians over the 1961–2010 period

| Romanian Carpathians regions | $SC_{max}$ (cm) | 90th percentile level (cm) |
|---|---|---|
| Eastern Carpathians | | |
| >800 m | 109 (Lăcăuţi, 1,776 m) | 107 |
| <800 m | 42 (Întorsura Buzăului, 707 m) | 31 |
| Southern Carpathians | | |
| >800 m | 145 (Vf. Omu, 2,504 m) | 121 |
| <800 m | 42 (Voineasa, 573 m) | 40 |
| Western Carpathians | | |
| >800 m | 126 (Vlădeasa 1,800 m, 1,836 m) | 121 |
| <800 m | 30 (Câmpeni, 591 m) | 30 |

Eastern Carpathians or only rarely, in some depressions of the Southern and Western Carpathians (e.g. Voineasa, Câmpeni).

The maximum snow depth ($SC_{max}$) marks the peak of the snow accumulation season, showing a high occurrence probability in early spring months (March–April interval), under favourable cyclonic flow conditions. $SC_{max}$ ranges between 24 and 109 cm in the Eastern Carpathians, 30 and 145 cm in the Southern Carpathians and 14 and 126 cm in the Western Carpathians. The extreme character of $SC_{max}$ is clearly evident from the regional histograms of daily snow depths (Fig. 7.40), as the maximum occurrence probability is attributed to daily snow depths less than 20 cm. Regionally, the highest snow depths exceeded 140 cm only in the alpine area of the Southern Carpathians (145 cm – Vf. Omu, 2,504 m a.s.l.), while in the other regions, it ranged between 109 cm (Lăcăuţi, 1,776 m a.s.l. – the Eastern Carpathians) and 126 cm (Semenic 1,436 m a.s.l. – the Western Carpathians) (Table 7.17). The peak $SC_{max}$ records over the study period were fairly comparable to the snow depths depicting the long-term 90th percentile level, derived from the daily manually-measured snow depths at the 30 weather sites located in the Romanian Carpathians. Relative to the 90th percentile level, daily snow accumulations higher than 50 cm are considered extreme for the areas above 800 m, showing an occurrence probability of 43 % across the region. Comparatively, in foothill areas, daily snow depths exceeding 15–20 cm are extreme.

### 7.5.7 *Snow Water Equivalent*

Seasonal snowpack and spring runoff are economically valuable water resources and key components of surface water balance. There are no comprehensive studies on the snow water equivalent regime (SWE) of the Romanian Carpathians. Ground measurements of SWE have been used mostly in hydrological studies focused on seasonal runoff estimations, connections between snow density, weather parameters and soil infiltration, the influence of vegetation cover on the snowmelt-runoff rate etc. (e.g. Grumăzescu and Stăncălie 1988; Copaciu and Tibacu 1993; Diaconu and Şerban 1994; Miţă and Stăncălie 1996, Miţă et al. 2003; Simota 2005). Valuable contributions to the study of the available snow water resources and its influencing factors were also provided by Stancălie (1991), Stăncălie et al. (2000, 2003, 2006).

The water storage accumulated as snow during winter is progressively or rapidly released in spring or early summer as melt water, under warmer and wetter weather conditions (rain-on-snow events). The peak monthly mean snow depths correspond largely to the highest snow water equivalent values (SWE), depicting the snow accumulation, peak accumulation and snowmelting seasons, whose intervals show obvious altitudinal differences across the Romanian Carpathians (Fig. 7.41):

– *The snow accumulation season* may last 6–8 months in alpine areas higher than 2,000 m. In these areas the snow accumulation season may onset the earliest, from August and may last until late March or early April. Below 2,000 m, the

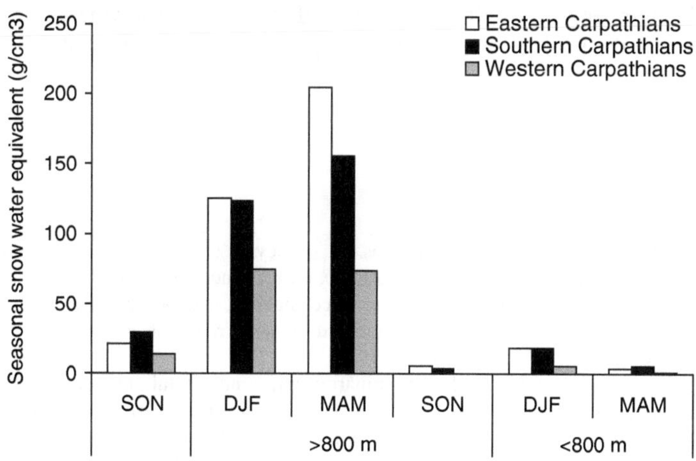


**Fig. 7.41** Altitudinal influence on the length of snow accumulation and snowmelt seasons across the Romanian Carpathians

season may start the earliest in late September to mid October but most frequently, from mid October, ending in February or March. In areas below 800 m, the water resources start to accumulate during the October–November interval, and exceptionally later in some intra-Carpathian depressions of the Western Carpathians. In this season, snowmelt events are very rare in this season at rates that do not exceed 10 cm/day. *The peak storage of available water resource* is typically in February (43.5 % cases) or March (30 % cases) and exceptionally in April (e.g. Vf. Omu station). In some areas above 1,500 m of the Southern Carpathians (Vf. Omu and Parâng), a secondary winter SWE peak was observed in December, contributing with 15–26 % to the annual snow water resource.

– *The snowmelt season* is visibly shorter than the accumulation one, its duration being highly dependent on elevation: 4–5 months in high-elevation areas above 2,000 m and 3–4 months in the mountain areas below 1,500–1,600 m. The onset of snowmelt is linked to the enhancement of solar radiation under more reduced cloudiness and implicitly, to the increasing frequency of thawing in the diurnal fluctuations of surface air temperature. The daily oscillations of snow depths of at least 20 cm may provide good evidence of rapid snowmelts, in response to the positive daytime and/or night time temperatures, frequently associated with liquid precipitation events. Micu (2009) investigated the occurrence of rapid snowmelt events over the 1961–2003 period, in the areas above 1,000 m of the Romanian Carpathians, by considering the daily snow depth decreases of at least 20 cm. Such events appeared to have the highest occurrence probability during the March–May interval (up to 26 % in March, 33 % in April and 19 % in May), lasting no more than three consecutive days. The advection of warm moist air in spring, under westerly, southerly or south-westerly circulations, was found to explain the temperature increases (e.g. the maximum temperatures values were in general higher than 5 or even 8 °C during the rapid snowmelt events) and the rain-on-snow events in relation to the snow melting. However, the positive temperature oscillations explained most of the snowmelt events identified across the Romanian Carpathians (up to 80 %), regardless of elevation.

## References

Ancey C, Charlier C (1998) Les avalanches. In: Ancey C (ed) Guide Neige et Avalanches: Connaissances, Pratiques, Sécurité. Édisud, Aix-en-Provence, pp 87–126

Andrițoiu N (1962) Calculul radiației directe și considerații asupra regimului radiativ în RPR (Determination of the direct solar radiation and the solar radiation regime in Romania). Cul Lucr Meteor/1960:171–182 (in Romanian)

Andrițoiu N, Ciocoiu I (1971) Unele determinări asupra gradientului vertical al radiației solare directe în regiunile de munte (Determination of the vertical gradient of the direct solar radiation in mountain regions). Cul Lucr Meteor/1968:181–193 (in Romanian)

Annual Bulletin on the Climate in WMO Region VI – Europe and Middle East (2000, 2007, 2010) http://www.dwd.de/rcc-cm

Apăvăloaie M (1972) Diferenţieri locale în regimul şi repartiţia precipitaţiilor atmosferice determinate de altitudine în sectorul mijlociu al văii Bistriţa (Local differentiations in the precipitation regime and its spatial distribution induced by height in the upper Bistriţa Valley). Lucr Simpoz Geogr Fiz Carpat 1972:253–261 (in Romanian)

Armstrong RL, Brun E (2008) Snow and climate: physical processes. Surface energy exchange and modeling. Cambridge University Press, Cambridge, p 256

Bacinschi D, Neacşa O (1985) Principalele caracteristici climatice ale Carpaţilor Româneşti (The main climatic characteristics of the Romanian Carpathians). Lucr Şt ICPCP–Braşov X:33–75 (in Romanian)

Bădescu V (1999) Correlations to estimate monthly mean daily solar global irradiation: application to Romania. Energy 24(10):883–893

Barbu I (1979) Cercetări privind influenţa reliefului în producerea rupturilor şi doborâturilor de zăpadă din aprilie 1977 în pădurile din Bucovina (Research related to the relief influence on the windthrow and snow-induced forest damages of April 1977 in Bucovina' forests). In: Proceedings of the national Conference for soil science. Braşov (in Romanian)

Barbu I (1982) Zonarea pe baze geomorfologice a vulnerabilităţii la zăpadă a arboretelor (Geomorphic-based zonation of forest vulnerabilty to snow-load). BSRG. Serie nouă. VI (LXXVI) (in Romanian)

Barbu I (1998) Propuneri de zonare teritorială a României în raport cu riscul apariţiei vătămărilor produse de zăpadă în păduri (Proposals of territorial zonation of snow-induced forest damages in Romania). Bucovina Forestieră 1–2:29–37 (in Romanian)

Barbu I (2004) Metode de evaluare a riscului de apariţie a vătămărilor de zăpădă în pădurile României (Methods for assessing the risk of snow-induced forest damages in Romanian forests). Bucovina Forestieră XII:1–2 (in Romanian)

Barry RG (2008) Mountain weather and climate, 3rd edn. Cambridge University Press, Cambridge, p 506

Bâzâc G (1981) Zone cu priorităţi pentru valorificarea potenţialului energiei eoliene în România (Prioritary areas for aeolian energy use in Romania). Hidrotehnica 26(8):225–227

Bâzâc G (1983) Influenţa reliefului asupra principalelor caracteristici ale climei României (Relief influence on the main climatic characteristics of Romania). Edit. Academiei Române. Bucureşti. p 179 (in Romanian)

Beşleagă N (1979) Aerologie şi meteorologie aplicată (Aerology and applied meteorology). Edit. Didactică şi Pedagogică. Bucureşti. p 223 (in Romania)

Beyer HG, Costanzo C, Heinemann D (1996) Modifications of the Heliosat procedure for irradiance estimates from satellite images. Sol Energy 56:207–212

Bogdan O (1978) Fenomene climatice de iarnă şi de vară (Winter and summer climatic phenomena). Edit. Ştiinţifică şi Enciclopedică. Bucureşti. p 120 (in Romanian)

Bogdan O (1985) Particularităţi ale calmului atmosferic din unele regiuni depresionare şi culoare de vale (Particularities of atmospheric calm in some depression areas and valley corridors). SCGGG-Geogr XXXII:14–19 (in Romanian)

Bogdan O (1993) Foehnul carpatic (Carpathian foehn). Anal Univ din Oradea-Geogr III:58–63 (in Romanian)

Bogdan O (1994) Culoarul Oltului – o "axă" de dicontinuitate topoclimatică (The Olt Corridor – a topoclimatic discontinuity axis). Anal Univ din Oradea-Geogr IV:100–108 (in Romanian)

Bogdan O (ed) (2008) Carpaţii Meridionali. Clima şi hazardele meteo-climatice şi impactul lor asupra turismului (The Southern Carpathians. Climate and meteorological-climatological hazards and their impact on tourism). Universităţii Lucian Blaga, Sibiu, p 324 (in Romanian)

Bogdan O, Coşconea M (2010) Riscul doborâturilor de arbori în România – cauzele (Windthrow risk in Romania – the causes). Riscuri şi Catastrofe 8(1):89–102, IX (in Romanian)

Bogdan O, Dragotă C (1997) Calmul atmosferic în România (Atmospheric calm in Romania). Anal Univ din Oradea-Geogr VII:96–104 (in Romanian)

Bogdan O, Niculescu E (1990) Un caz tipic de foehn în România (A typical foehn case study in Romania). SCGGG–Geogr XXXVII:95–103 (in Romanian)

Bogdan O, Marinică I, Mic L (2011) Characteristics of the 2007 canicular summer in Romania. Rev Roum de Géogr 55(1):45–56

Bordei-Ion E (1983) Rolul lanţului alpino-carpatic în evoluţia ciclonilor mediteraneeni (The role of the Alpine-Carpathian Chain in the evolution of Mediterannean cyclones). Edit. Academiei Române. p 136 (in Romanian)

Bordei-Ion N (2008) Fenomene meteo-climatice induse de configuraţia Carpaţilor în Câmpia Română (Meteorological-climatological phenomena induced by the configuration of the Carpathians in the Romanian Plain region). Edit. Academiei Române. Bucureşti. p 176 (in Romanian)

Buiuc M (1984) Estimarea radiaţiei solare pe teritoriul României (Estimation of solar radiation in Romania). SC-Fundamentarea meteorologică şi hidrologică a resurselor energetice neconvenţionale. INMH. Bucureşti (in Romanian)

Cano D, Monget JM, Albuisson M, Guillard H, Regas N, Wald L (1986) A method for the determination of the global solar radiation from meteorological satellite data. Sol Energy 37:31–39

Cheval S, Baciu M, Dumitrescu A, Breza T, Legates DR, Chendeş V (2010) Climatologic adjustments to monthly precipitation in Romania. Int J Climatol 31:704–714

Ciocoiu I (1976) Studiul resurselor de energie solară pe teritoriul RSR (mss.) (The study of solar energy resources of Romania). IMH. Bucureşti. p 60 (in Romanian)

Ciulache S (2002) Meteorologie şi climatologie (Meteorology and climatology). Edit. Universitară. Bucureşti. p 469 (in Romanian)

Clima României (2008) (The climate of Romania). Edit Academiei. Bucureşti (in Romanian)

CM-SAF Surface Radiation MVIRI Data Set 1.0 – Monthly Means/Daily Means/Hourly Means. doi: 10.5676/EUM_SAF_CM/RAD_MVIRI/V001

Copaciu V, Tibacu L (1993) Correlation between the density of the snow cover and meteorological parameters. In Proceedings of the international symposium on precipitation and evaporation, Slovak Hydrometeorological Institute, Bratislava: 323–328

Cortemiglia GC, Rossetti R, Ghezzi A, Tortelli M (1989) Variations thermiques avec l'altitude dans les Alpes occidentales italiennes, Actes du colloque de Pavia, Italie, 1–3 Juin 1989. Publ Assoc Int Climatol 2:19–26

De Saintignon MF (1986) Décroissance de la température avec l'altitude dans le Jura francais en fonction du site des postes de mesure: Comparison avec les Alpes et les Vosges, Proc. Int. Symp. on Topoclimatology and its Applications., Liège. Mont-Rigi. Belgium. IGU: 1–11

Diaconescu G (1966) Mersul radiaţiei solare globale şi fenomenele meteorologice (Variation of global solar radiation and the meteorological phenomena). Cul Lucr Meteor 1964:125–134 (in Romanian)

Diaconu C, Şerban P (1994) Sinteze şi regionalizări hidrologice (Hydrological syntheses and regionalizations). Edit. Tehnică. Bucureşti (in Romanian)

Douguédroit A, De Saintignon MF (1970) Méthode d'étude de la décroissance des températures en montagne de latitude moyenne: Exemple des Alpes françaises du Sud. Rev Geogr Alp 58:453–472

Douguédroit A, De Saintignon MF (1981) Décroissance des températures mensuelles et annuelles avec l'altitude dans les Alpes du Sud et en Provence (séries 1959–1978). Eaux et Climat: Mélanges Géographiques Offerts en Homage à Charles-Pierre Péguy, ER 30:179–194

Dragotă CS (2006) Precipitaţiile excedentare din România (Excedentary precipitation in Romania). Edit. Academiei Române. Bucureşti. P 176 (in Romanian)

Duchon ED, Essenberg GR (2001) Comparative rainfall observations from pit and aboveground rain gauges with and without wind shields. Wat Resour Res 37(12):3253–3263

EEA (2004) Arctic environment: European perspectives. Why should Europe care? Environmental issue report no. 38. p 60

Elder K, Dozier J, Michaelsen J (1991) Snow accumulation and distribution in an alpine watershed. Water Resour Res 27:1541–1552

Esteban P, Jones PD, Martin-Vide J, Mases M (2005) Atmospheric circulation patterns related to heavy snowfall days in Andorra, Pyrenees. Int J Climatol 25:319–329

Etienne C, Lehmann A, Lopez-Moreno JI, Goyette S, Beniston M (2010) Spatial predictions of extreme wind speeds over Switzerland using generalized additive models. J Appl Meteorol 49:1956–1970

Fărcaş I (1981) Contributions au problème de la variation annuelle de la radiation solaire directe sur le surfaces avec différentes orientations et inclinations aux latitudes moyennes. RRGGG-Géogr 25(2):67–72

Florescu I (1981) Silvicultură (Forestry). Edit. Didactică şi Pedagogică. Bucureşti. p 294 (in Romanian)

Gaceu O (2005) Clima şi riscurile climatice din Munţii Bihor şi Vlădeasa (The climate and climatic risks of the Bihor and Vlădeasa Mountains). Edit. Universităţii din Oradea, p 284 (in Romanian)

Grumăzescu H, Stăncălie G (1988) The surveillance of snow cover dynamics and the forecast of the snowmelt runoff in the Carpathian Basin using remote sensing data. In: Proceedings of the international workshop on hydrology of mountain areas, Strbske Pleso, Cehoslovacia

Gugiuman I, Stoian R (1972) Zăpada în Carpaţii Româneşti (Snowpack in the Romanian Carpathians). Lucr Simpoz Geogr Fiz Carpat 1970: 283–296 (in Romanian)

Hammer A, Heinemann D, Hoyer C, Kuhlemann R, Lorenz E, Müller R, Beyer HG (2003) Solar energy assessment using remote sensing technologies. Remote Sens Environ 86:423–432

Haring F, Iuga M (1970) Cercetări privind rupturile de zăpadă din arboretele de molid din Munţii Maramureşului (Research on the snow-induced damages in the spruce forests of the Maramureş Mountains). Analele ICAS 27(2):479–500 (in Romanian)

Hess M (1965) Vertical climatic zones in the Polish Western Carpathians. Zeszyty Naukowe UJ Prace Geograficzne 11:1–267

Hess M (1971) Vertical climatic zones in north and south Carpathians and their thermic characteristics. Folia Geogr Ser, Geogr Phy Kraków 5:15–23

Holtmeier F-K (2009) Mountain timberlines. Ecology, patchiness, and dynamics – advances in global change research 36. Springer, Dordrecht/Boston/London

Ichim R, Barbu I (1981) Rupturile şi doborâturile de zăpadă din pădurile judeţului Suceava (Snow-induced forest damages in the Suceava County). Seria II. ICAS. Bucureşti (in Romanian)

Ion-Bordei N (2008) Fenomene meteoclimatice induse de configuraţia Carpaţilor în Câmpia Română (Meteorological-climatic phenomena induced by the configuration of the Carpathians in the Romanian Plain). Editura Academiei Române, Bucureşti, p 174 (in Romanian)

Kocin PJ, Uccellini LW (1990) Snowstorms along the Northeastern Coast of the United States: 1955 to 1985. Meteorol Z 56:418–428

Landolt E (1983) Probleme der Hohenstufen in den Alpen. Bot Helv 93:255–268

Macovei C (2009) The wind – climatic risk factor for Suceava's county forestry stock. Present Environ Sustain Dev 3:275–286

Mani MS (1968) Ecology and biogeography of high-altitude insects. Junk, The Hague, p 527

Marcu G, Stoica C, Beşleagă N, Stoian R, Ceianu I, Dissescu R, Pavelescu I (1969) Doborâturile de vânt în anii 1964–1966 în pădurile din România (Windthrows of the 1964–1966 in the Romanian forests). Edit. Agrosilvică. Bucureşti. p 161 (in Romanian)

Martonne E de (1907) Recherches sur l'évolution morphologiques des Alpes de Transylvanie. Paris

McClung DM, Scharer P (eds) (2006) The avalanche handbook, 3rd edn. The Mountaineers Books, Seattle, p 342

Micu D (2009) Snow pack in the Romanian Carpathians under changing climatic conditions. Meteorol Atmos Phys 105(1–2):1–16

Micu D, Mărăşoiu D (2009) Ninsorile abundente din Carpaţii Meridionali (Heavy snowfalls in the Southern Carpathians). Revista Geografică. XVI. Serie nouă: 5–14 (in Romanian)

Mihai E (1975) Frecvenţa inversiunilor de temperatură în Depresiunea Braşov (Frequency of temperature inversions in the Braşov Depression). SCGGG-Geogr XVI (2):205–212 (in Romanian)

Miţă P, Stăncălie G (1996) Methods of computing of the variation of the characteristics of the snow cover and runoff generated by snowmelting. Centre d'Études et des Recherches Eco-Géographique (CEREG), Strasbourg

Miţă P, Simota M, Ungureanu V, Mătreaţă S (2003) Floods of combined origin – rainfall and snowmelt – in Romanian small catchments. In: Proceedings of the 9th ERB conference, Slovakia, September 25–28, 2002, IHP-VI Technical Documents in Hydrology, p. 67

Mortensen NG, Petersen EL (1998) Influence of topography input data on the accuracy of wind flow modeling in complex terrain. In: Proc. 1997 European Wind Energy Conf. Dublin, Ireland. Electricity Supply Board and the European Commission, pp 317–320

Mote TL, Gamble DW, Underwood SJ, Bentley ML (1997) Synoptic-scale features common to heavy snowstorms in the southeast United States. Weather Forecast 12:5–22

Moţoiu M (2008) Avalanşele şi impactul lor asupra mediului. Studii de caz din Carpaţii Meridionali (Snow avalanches and their impact on environment. Case studies from the Southern Carpathians). Edit. Proxima. Bucureşti. p 279 (in Romanian)

Muică C (1989) Vegetaţia – Potenţialul mediului în Subcarpaţii Buzăului (Vegetation – Environmental potential of the Buzău Subcarpthians). Institutul de Geografie. Bucureşti. pp 114–123

Neacşa O, Popovici C (1969) Repartiţia duratei de strălucire a Soarelui şi a radiaţiei globale pe teritoriul RSR (Distribution of sunshine duration and global solar radiation in Romania). Cul Lucr Meteor 1967 (in Romanian)

Neacşa O, Popovici C, Sabău A, Radu D, Susan V (1973) Viteza maximă a vântului cu diferite probabilităţi de producere în zona muntoasă degajată a Carpaţilor Româneşti (Maximum wind speed with different occurrence probabilities in the high-elevation areas of the Romanian Carpathians). Hidrotehnica 18:590–591 (in Romanian)

Neamu G, Teodoreanu E (1972) Repartiţia precipitaţiilor atmosferice în raport cu altitudinea în Carpaţii Româneşti (Precipitation distribution by height in the Romanian Carpathians). Lucr Simpoz Geogr Fiz Carpat 1972:275–283 (in Romanian)

Niculescu E (1993) Influenţa suprafeţei active montane româneşti asupra climatului şi topoclimatului de tip carpatic (Influence of the mountain underlying surface on the Carpathian climate and topoclimate). Analele Universităţii din Oradea-Geogr (III):64–68 (in Romanian)

Ohmura A, Dutton EG, Forgan B, Fröhlich C, Gilgen H, Hegner H, Heimo A, König-Langlo G, McArthur B, Müller G, Philipona R, Pinker R, Whitlock CH, Dehne K, Wild M (1998) Baseline Surface Radiation Network (BSRN/WCRP): new precision radiometry for climate research. Bull Am Meteorol Soc 79(10):2115–2136

Oprea C (2005) Climatul radiativ pe teritoriul României. O abordare geografică a radiaţiei solare (Radiative climate of Romania. A geographical approach of solar radiation). Book in digital format. ISBN: 973-0-03879. Bucureşti. p 250 (in Romanian)

Pardé M (1955) Fleuves et rivières. Colin, Paris, p 244

Parish R (2002) Mountain environments. Prentice Hall, New York, p 348

Popa I (2000) Sisteme de cartare a zonelor de risc la doborâturi produse de vânt (approaches for mapping the windthrow risk areas). Revista Pădurilor 4:35–41 (in Romanian)

Popa I (2005) Doborâturi produse de vânt – factor de risc în ecosistemele forestiere montane (Windthrow – risk factor in the mountain forest ecosystems). Anal ICAS 48 (in Romanian)

Posselt R, Mueller R, Stöckli R, Trentmann J (2012) Remote sensing of solar surface radiation for climate monitoring – the CM-SAF retrieval in international comparison. Remote Sens Environ 118:186–198

Rolland C (2003) Spatial and seasonal variations of air temperature lapse rates in Alpine regions. J Climate 16(7):1032–1046

Sevruk B (1982) Methods of correction for systematic error in point precipitation measurement for operational use. WMO Oper Hydrol Rep 21(589):91 p

Simota M (2005) Modeling the dynamics of snow cover during accumulation and melting periods for the forecasting of snow-melt runoff. In: Proceedings of the XXth conference of the Danube countries, Sept. 4–6, Bratislava, Slovakia

Şorodoc C (1962) Formarea şi evoluţia ciclonilor mediteraneeni şi influenţa lor asupra timpului în România (The formation and evolution of the Mediterranean cyclones and their influence on weather in Romania). Cul Lucr Meteor 1960: 5–41 (in Romanian)

Stan M (1950) Fluxul de insolaţie pe suprafeţe cu orientări şi înclinări diferite pentru latitudinile de 44–48° (The direct solar radiation flux on surfaces with different aspects and slopes for 44–48°N). Mem. şi Studii. III. 3. IMC (in Romanian)

Stăncălie G (1991) Remote sensing monitoring of snowpack dynamics in view of estimating the snowcover water resources in drainage basins of hydroelectric power interest. IUGG-XX General Assembly. Congress Report. Wien

Stăncălie G, Catană S, Alecu C (2000) Evaluation and monitoring of snowcover water resources in Carpathian basins by remote sensing and GIS techniques. In: Proceedings of the 26th international conference on alpine meteorology, Innsbruck, Austria

Stăncălie G, Iordache A, Catană S (2003) Study of the influences of geographical factors and meteorological parameters on the snow cover evolution in Romanian mountainous basins. ICAM-MAP Meeting 2003. Publications of MeteoSwiss, Brig, No.66

Stăncălie G, Catană S, Flueraru C (2006) Evaluation and monitoring of snow cover water resources in Carpathian basins using geographic information. In: Pollert J, Dedus B (eds) Security of water supply systems: from source to tap, NATO security through science series, vol. 8. Springer Netherlands, pp 125–135

Stăncescu I (1983) Carpaţii – factori modificatori ai climei (The Carpthians as modifier factors). Edit. Ştiinţifică şi Enciclopedică. Bucureşti, p 139 (in Romanian)

Stăncescu I, Damian D, Şerbu C (1986) Unele particularităţi în repartiţia precipitaţiilor atmosferice în zona Carpaţilor Româneşti şi a regiunilor limitrofe datorate proceselor foehnale (Particularities of precipitation distribution in the Romanian Carpathians and their limitrophe regions induced by foehn processes). Studii şi Cercetări-Meteor: 215–223 (in Romanian)

Stănescu VA, Ungureanu V (1997) Hydrological regimes in the FRIEND-AMHY area: space variability and stability. In: FRIEND '97 – Regional Hydrology: Concepts and Models for Sustainable Water Resource Management. IAHS Publ. n 246

Stoenescu ŞM (1951) Clima Bucegilor (The climate of the Bucegi Mountains), Edit. Tehnică, Bucureşti, p 219 (in Romanian)

Święchowicz J (2012) Climate and hydrology. In: Lóczy D, Stankoviansky M, Kotarba A (eds) Recent landform evolution. The Carpatho-Balkan-Dinaric Region. Springer Netherlands, pp 53–55

Ţâştea D (1961) Calculul şi repartiţia radiaţiei solare pe teritoriul RPR (Determination and spatial distribution of solar radiation in Romania). MHGA.1 (in Romanian)

Ţâştea D, Lorentz R (1985) Stabilirea corelaţiilor dintre gradientul baric şi vânturile puternice produse în diverse zone ale României (Correlations between the pressure gradient and gusty winds in different areas of Romania). Cul Lucr Meteor 1982 (in Romanian)

Ţâştea D, Neacşa O (1974) Variaţia altitudinală a principalilor parametri climatici în zona Bâlea-Capra din Masivul Făgăraş (Vertical zonation of the main climatic parameters in the Bâlea-Capra area, the Făgăraş Mountains). Studii de climatologie 1:105–209, in Romanian

Ţâştea D, Călinescu G, Soare E (1990) Consideraţii privind diferenţa de presiune atmosferică dintre sudul şi nordul Culoarului Oltului (Carpaţii Meridionali) ca factor generator al intensificării vânturilor din sectorul sudic al staţiei meteorologice Boiţa (Considerations on the atmospheric pressure differences between the southern and northern parts of the Olt Corridor – the Southern Carpathians –, as genetic factors of wind intensifications in the southern sector of Boiţa weather station). Studii şi Cercetări-Meteor, 4 (in Romanian)

Teodoreanu E (1980) Culoarul Rucăr-Bran. Studiu climatic şi topoclimatic (The Rucăr-Bran Corridor. A climatic and topoclimatic study). Edit. Academiei. Bucureşti. p 166 (in Romanian)

Tilinca Z, Fărcaş I, Mihăilescu M (1976) Contribuţie la studiul sinoptic al foehnului în Munţii Apuseni (Contribution to the synoptic study of foehn in the Apuseni Mountains). Studii şi Cercetări-Meteor I(2):567–584 (in Romanian)

Urdea P (2012) Periglacial processes. In: Lóczy D, Stankoviansky M, Kotarba A (eds) Recent landform evolution. The Carpatho-Balkan-Dinaric Region. Springer Netherlands, pp 266–269

Vespremeanu-Stroe A, Cheval S, Tatui F (2012) The wind regime of Romania – characteristics, trends and North Atlantic oscillation influences. Forum Geogr 2:118–126, XI

Voiculescu M (2002) Fenomene geografice de risc în Masivul Făgăraş (Geographical risk phenomena in the Făgăraş Massif). Edit. Brumar. Timişoara. p 231

Voiculescu M (2009) Snow avalanche hazards in the Făgăraş Massif (Southern Carpathians): Romanian Carpathians-Management and perspectives. Nat Hazards 51(3):459–475

Voiculescu M, Onaca A, Milian N, Ardelean F, Török-Oance M, Stăncescu M (2010) Analysis of snow avalanche from mars 07, 2007 within the Călţun-Negoiu area, in the Făgăraş massif (Southern Carpathians). Anal Univ din Oradea-Geogr XX(1):22–33

Whiteman CD (2000) Mountain meteorology. Fundamentals and applications. Oxford University Press, Oxford, p 355

Wild R, O'Hare G, Wilby R (1996) A historical record of blizzards/major snow events in the British Isle, 1880–1989. Weather 51:81–90

Winstral A, Elder K, Davis RE (2002) Spatial snow modeling of wind-redistributed snow using terrain-based parameters. J Hydrometeorol 3:524–538

WMO (World Meteorological Organization) (2009) Guidelines on analysis of extremes in a changing climate in support of informed decisions for adaptation, Climate data and monitoring WCDMP-No. 72: 55 p

Zăvoianu I, Alexandrescu M, Anghel C, Mustăţea A (1995) Corelaţii între precipitaţii şi altitudine în Carpaţii Meridionali (Correlations between precipitation and height in the Southern Carpathians). Studii şi Cercetări-Geogr XLII:13–19 (in Romanian)

# Chapter 8
# Observed Variability and Trends from Instrumental Records

**Abstract** This chapter analyses the recent climatic trends in the Romanian Carpathians, with an accent on seasonal changes. Temperature trends are increasing in winter, spring and summer, while they are completely absent in autumn, which is the single stable season from a thermal point of view. On the other hand, autumn is the only season when significant increasing trends in precipitation have been found. The average wind speed is decreasing in all seasons, confirming most of the findings regarding the general tendency in the Northern Hemisphere. Snow-related trend analysis shows general decreasing trends in mean snow depth, number of days with snow cover, number of days with snowfall, and continuous snow cover duration. The increase in temperature at most of the locations, together with the slight decrease in winter precipitation explains the reduction of the snowfall days. The number of snowfall days, snow duration and mean snow depth present strong negative correlations with the NAO index for the same period (DJF). The large-scale circulation over the North Atlantic has a considerable effect on the winter season in the Romanian Carpathians.

Variability is visible in climatic oscillations relative to the normal state, conventionally established for a sufficiently long period of observations and deemed to be climatologically representative. Considering the present-day climate, the climatological baseline established by the World Meteorological Organization (WMO) is based on the 30-year average calculated for the 1961–1990 interval, which is marked by an intensification of human activity (Stehr and von Storch 2010). At present, there is evidence of a progressive climate change from observational datasets worldwide, suggesting visible modifications in the average values or in the overall variability of climate characteristics within a given area. Busuioc et al. (2010) emphasizes that these climatic changes are statistically outlined for time periods of different lengths, but covering at least 30 years (the WMO standard). This chapter presents the analysis of climate variability and associated trends in the Romanian Carpathians (above the 1,000 m altitude), over a 50-year period of observations (1961–2010).

According to the 2007 and 2013 reports of the Intergovernmental Panel of Climate Change (IPCC 2007, 2013), the global climate evolution over the last hundred years shows a clear and ongoing warming process. The decadal warming



D.M. Micu et al., *Climate of the Romanian Carpathians*,
Springer Atmospheric Sciences, DOI 10.1007/978-3-319-02886-6_8

rate (0.13 °C/decade) registered in the last 50 years was twice that of the last century (0.07 °C/decade). The warming process proved to have greater intensity in Europe than on a global scale. In some European areas, the annual average temperature rose by 1.2 °C, with a decadal warming rate of only 0.2 °C during 1996–2007. Most seasons (summer, winter and spring) are significantly affected by warming, an exception being autumn, which shows no specific trends. The European regions that are mainly in question are the Iberian Peninsula, central and northeastern Europe, but also some mountain regions which proved to be quite sensitive to positive temperature oscillations (EEA 2008, 2010; Böhm et al. 2001; Fagre et al. 2003; Klein Tank 2004).

Over 1900–2005 there was an extremely great global spatial and temporal variability of precipitation: some regions were under a increasing trend (e.g. the eastern part of North and South America and the north of Europe) while others, under a decreasing trend (e.g. the Sahel, the Mediterranean region of Europe, the south of Africa and some regions in the south of Asia) (IPCC 2007; Beniston and Rebetez 1997). Simultaneously, droughts became more severe and long lasting, mostly in the tropical and extra-tropical regions and particularly after 1970. Temperature increases triggered prevailing liquid precipitation rather than snow in the cold season, with earlier occurrences in spring (Cayan et al. 2001; Easterling 2002; Groisman et al. 2005; IPCC 2007). The same precipitation variability patterns were underlined from land observations in many European regions (EEA 2008, 2010).

An important feedback mechanism of the climate system is the snowpack, which is highly sensitive both to temperature and precipitation fluctuations, being considered a reliable indicator of climate change. Its spatial expansion reveals the direct influence of thermal and radiative variations and a key evidence of the local and regional effects of the general climate warming. Over 30 % of the terrestrial continental surface is seasonally covered with snow, and approximately 50 % in the Northern Hemisphere, mostly in January and February (46 mill. $km^2$) (Armstrong and Brodzik 2001). As the climate was warming up, the Northern Hemisphere snow-covered surface decreased by some 10 %, snowpack-free intervals increasing by 8.8 days/decade, especially in the 45–75°N latitude belt (1971–1994) (Dye 2002). In 1966–2005, the European part of the Northern Hemisphere registered a decrease of snow cover area at a rate of 1.3 %/decade (EEA 2008). The last 2007 and 2013 IPCC reports acknowledge a great inter-annual variability of snow cover, mostly in autumn (absolute values) and summer (relative values). From 1970, the Northern Hemisphere snow cover area was reduced substantially, particularly in spring and summer, less so in winter, although warming trends in the cold season are considerable. The observed trends over the second half of the twentieth century showed earlier snowmelt in spring, mostly between 1972 and 2000. This snow regime change signal is already visible in the European Alpine region, where the snow season tends to shrink, mainly due to earlier spring snowmelt rather than late snowfalls (Laternser and Schneebeli 2003).

Mountain regions are among the most fragile natural environments on Earth (Diaz et al. 2003). Their overwhelming importance resides in the huge variety of ecosystems, biodiversity, water resources volume and ecosystem services provided

(Messerli and Ives 1997; Woodwell 2004), whose influence prevails beyond their geographical boundaries. These regions are extremely sensitive and vulnerable to a wide range of pressures induced by population and economic growth (Spehn et al. 2002). However, the recent climate evolution observed in many mountain regions worldwide (e.g. decreases of snow cover area, alteration of liquid and solid falls patterns, more frequent heat waves, accelerated ice loss by glacier melt, permafrost decline etc.) poses new threats to these sensitive environments by affecting the frequency and severity of multiple non-seismic hazards (e.g. storms, snow avalanches, landslides). Numerous studies have stressed the great sensitivity of mountain regions to the recently observed global climate variability that generally affects the cryosphere components (glacier, snowpack and permafrost), water resources and hydrological cycle, timberline height and the vertical distribution of vegetation species (e.g. Barnett et al. 2005; Beniston et al. 2003, 1997; Dyurgerov 2003; Grabherr 2003; Wilson et al. 2005; Schröter et al. 2005; Nogués-Bravo et al. 2007; Bavay et al. 2009 etc.).

In general, Europe has shown a greater warming trend since 1979 compared to the global mean and the related climate trends in mountainous regions are even more pronounced (Böhm et al. 2001). Mountain regions have been designated as 'early indicators of climate change' [UNEP Conference, April 2008, Padova]. The observed changes in the temperature, precipitation and snow regimes, in the most important mountain ranges of Europe over the twentieth and the early twenty-first centuries, are shown in the Table 8.1. The Romanian Carpathians, though a component of the European Alpine system, were less investigated previously as a whole. This study helps complete the picture of general regional (and local) climate change effects within this mountain region in the areas below and above 800 m. This section describes the variability signals over 1961–2010 in the regime of four essential variables of the Carpathians climate, which are subject to most changes induced by global climate warming (air temperature, precipitation, wind and snowpack)

## 8.1 Air Temperature

The most significant temperature increase is obvious in the mountain areas of less than 1,800 m altitude (from 0.15 to 0.28 °C/decade). The intensity of warming is showing small differences at regional scale, rather than in altitude. High-elevation areas (above 1,800 m) experienced a less intense warming compared to the lower elevation ones, below 1,700 m. This signal is significant at the 10 % confidence level, particularly if referring to the annual trends of minimum temperatures ($T_{min}$). Exceptionally, in some areas around 1,300–1,500 m of the Southern Carpathians (at Sinaia 1,500 m, Parâng and Păltiniş stations), the maximum temperature ($T_{max}$) rise is stronger, at a rate of 0.19–0.27 °C/decade (Fig. 8.1). A clear difference in the behavior of $T_{min}$ and $T_{max}$ that depends on altitude and location was also observed in Switzerland, including the mountain regions (Jungo and Beniston 2001).

**Table 8.1** Regional signals of climate variability derived from in situ observations in the most important mountain ranges of Europe

| Mountain units | Observed changes in temperature regime | Observed changes in precipitation regime | Observed changes in snowpack regime | References |
|---|---|---|---|---|
| French Alps | The general warming rate ranges between 0.02 and 0.04 °C/year. Mean annual temperature increase of 0.9 °C (1901–2000); temperature increase is more pronounced during the day than during the night (e.g. warming rate by maximum temperature values is about 0.9–1.1 °C/100 year). Seasonally, in the Écrins (Monétier-les-Bains station, 1,490 m) and Dévoluy Massifs (La Salette station, 1,770 m) warming process became more visible after 1960, particularly during summer (0.9 °C/100 years), winter 0.7 °C/100 years) and autumn (0.6 °C/100 years). The annual number of freezing decreased (12–14 % after 1980). The alpine climate of Oisans (Sain Chris-tophe, 1,570 m) and Briançonnais Massifs (Southern French Alps) became warmer, particularly during the 1990s | The regional precipitation variability signals are unevenly distributed spatially. Over the 1984–1999 period, in the northwestern part of the French Alps the observed variability patterns showed a general increase of annual precipitation amount (also associated to an increase of the annual precipitation maximum). The southeastern part of the Alps showed a notable decrease. The precipitation regime of the Southern Alps (Écrins and Devoluy massifs) shows a slightly increasing trend of precipitation levels in spring, summer and autumn (1901–2000), while winter is on a slight decrease. The inter-annual precipitation variability is high also in the Maritime Alps. In terms of precipitation extremes, there is a reduction in the number of consecutive dry days in spring and an increase in autumn, particularly in the southern Alps, while in the northern Alps there is an increase in the frequency of | Snowpack regime reveals a strong inter-annual variability of its related parameters. The main climate change signal in the region suggests a decrease of snow depth and snow cover area over the average snow season, most visible over the last decades of the twentieth century. Even after 1990, the French Alps were affected by a lack of snow and snowfalls, the multiannual variability has not revealed a significant trend of the annual average snow depth, snowfall days and snow extremes. Sensitivity studies show that at lower elevations (e.g. below 1,500 m) snow regime is extremely sensitive to small changes in temperature, especially in the southern part of the French Alps. Changes in the precipitation amount also influence the maximum snow depth (or snow water equivalent) much more than snow cover duration | Martin et al. (2001)<br>Etchevers and Martin (2002)<br>Bárdossy et al. (2003)<br>Jomelli et al. (2004)<br>Martin and Etchevers (2005)<br>ONEREC (2007)<br>Bodin et al. (2009) |

| | | | | |
|---|---|---|---|---|
| | | heavy rainfalls, in winter, spring and autumn. In summer, the number of heavy rainfall days (≥30 mm) is on the increase in the Écrins Massif even if the annual precipitation amount does not display a statistically significant trend after 1980 | | |
| German Alps | During the twentieth century mean annual temperature increased 0.5–1.2 °C. Seasonally, winter and summer became particularly warmer. Over the year the warming process is notable in January, February, March and October, but the warming rates are the highest in August (0.7–1.7 °C) and December (1.8 and 2.7 °C) | Over the twentieth century, the precipitation regime displayed different variability patterns. Hence, while the pre-alpine region experienced a slight precipitation decrease, the high alpine areas are marked by a positive trend in precipitation amounts, particularly in winter. Seasonally, winter (particularly in its end part) and spring show a 20–30 % increase of precipitation amounts, while in summer the precipitation deficit is notable (20 % decrease) over the entire southern part of Germany (including the alpine region). The precipitation regime variability also shows an intensification of precipitation extremes, particularly those suggesting an increase of rainfall intensity. Over 1958–2001, the Germen Alps experienced an increase of heavy rainfalls at a rate of 20–30 % | The snow cover duration in the southern part of Germany is getting shorter (the reduction of the number of snow cover days is statistically significant, at a rate of 10–20 % in the mentioned region, and of less than 10 % in the high alpine areas). This trend is blurred by altitude | Hennegriff et al. (2006), Seiler (2006) |

(continued)

**Table 8.1** (continued)

| Mountain units | Observed changes in temperature regime | Observed changes in precipitation regime | Observed changes in snowpack regime | References |
|---|---|---|---|---|
| Italian Alps | The signal of a warmer mountain climate is observable at seasonal time-scale, by mean and extreme daily temperature values. At the I of twenty-first century (during the 2003–2006 interval), temperature anomalies recorded at above 2,400 m (Lago Valsoera station, located at 2,444 m in the Grand Paradiso Massif, northern Italy) reached 1.8 °C against the 1959–2002 mean. The winter of 2006/2007 was the mildest in the last two centuries for the entire alpine regions of northern Italy, when the anomalies reached 2.6 °C against the winter temperature mean of the 1961–1990 reference period (Dormais station) | The observed trends in the precipitation regime variability are very different in terms of their spatial distribution and statistical significance (1958–2001). Most of the visible changes are coming from the precipitation extremes. The northern Italian Alps are marked by a reduction of heavy rainfalls frequency, particularly in winter and spring and an increase during summer. Seasonally, the maximum number of consecutive dry days has notably increased only in winter, while there are no significant changes in their variation. The 50-year period of observations made in the western part of the Alps are not drawing significant variability signals in precipitation regime, most of the changes range from −0.05 to +0.06 % (as Mann-Kendall and Monte Carlo statistics showed). However, in this alpine region there is a major concern in terms the length of dry intervals, especially after 1989, emphasizing a trend towards more persistent dry (droughty) intervals | The change signals in the number of snow cover days in the alpine Italian region varies by decade: the 1971–1986 period featured and alternation of extremely snowy winters (e.g. 1977, 1978) and less snowy winters (e.g. 1973, 1981); after 1990, winters show a general trend towards a lack of snow. This trend maintains at the beginning of the twenty-first century (e.g. during the 2003–2006 interval the snow accumulation was 40 % lower than the 1959–2002 average). This precipitation change signals in the Southern Alps (Italian Alps) conform to observed trends in the precipitation regime of the Eastern and Central Alps (Swiss and German Alps) but they are slightly different than those observed in the Western Alps (French Alps) | Bárdossy et al. (2003), Beniston (2005), Valt et al. (2005), Ciccarelli et al. (2008). |

| | | | | |
|---|---|---|---|---|
| Swiss Alps | The general warming rate computed over the 1900–2006 period is about 1.47 °C. In some alpine areas, temperature increase was even more pronounced (e.g. at Saentis station located 2,500 m – alpine climate became milder at a rate of 3.8 °C/100 years). In these areas the warming process has intensified since the 1990s. The temperature change signals at 18 high-elevation sites are visible since the end of the 1970s and after, 1980. The alpine climate tends to become warmer, particularly during the night (by minimum temperature values) than during the day | The most visible changes in precipitation regime were observed on a seasonal scale, particularly in winter, when precipitation amounts show a statistically significant increase, at a rate of 15–20 % (1901–1994). At high alpine sites (Saentis, station, 2,500 m), there is a 3.3 mm/day increase in winter precipitation (1961–1990). The frequency of heavy rainfalls increases, particularly in summer (mostly visible in the southern alpine areas), due to the stronger thermal convection processes and the higher frequency of unstable weather episodes (associated with an intensified cyclonic activity). Generally, winters are tending to become milder, by temperature values, but also in terms of a greater frequency of liquid falls, particularly in the mountain areas below 1,700-2,000 m (above this altitude solid falls are still dominant) | A general transition from cold and snowy winters to milder winters with more liquid falls was observed in the second half of twentieth century. The number of snowfall days is on a notable decrease, this signal being obvious particularly in mountain areas below 1,300 m altitude. The seasonal snow depth and the number of snow depth larger than 20 cm indicates a considerable variability over the last century, less visible in its first half but significant in its second half. In some high alpine areas (above 2,000 m), the snow accumulation process maintains its characteristics or is even tending to intensify (Saentis, +35 % increase in seasonal snow depth in respect to the 1961–1990 period). Referring to snow extremes (1-day and 3-days cumulated snow depths) the change signals are rather weak at seasonal scale, most of the changes being obscured by the year-to-year variability | Keller (2000), Keller et al. (2000), Beniston et al. (2003), Beniston (2005), Frei and Schär (2001), Laternser and Schneebeli (2003), Schneebeli et al. (1997) |
| | The extreme heat wave of the summer of 2003 induced lower maximum temperature anomalies than those recorded during winter in the second half of the twentieth century (the last two to three decades): the average of winter maximum temperature exceeded 15 °C at Saentis (2,500 m), Grand-Saint-Bernard (2,479 m) and Jungfraujoch (3,572 m) weather stations. The warming process is generally more intense in winter than in | | | |

(continued)

**Table 8.1** (continued)

| Mountain units | Observed changes in temperature regime | Observed changes in precipitation regime | Observed changes in snowpack regime | References |
|---|---|---|---|---|
| | other seasons due to the higher frequency, intensity (e.g. cases of +10 °C exceedance of daily maximum temperatures have increased notably after 1990, up to 20 cases/winter) and duration of heat waves | | | |
| Tatra mountains | The high-mountain area of the Tatras (above 1,900 m) experienced a temperature increase at a rate of 0.02 °C/year (1966–2006). The warming is the highest and statistically significant during summer (0.04 °C/year), while in winter the rates are slightly lower (0.02 °C/year). The increase of maximum temperature is slightly higher than that of the minima, during the whole year and particularly in summer (0.02 °C/year). The warming signals are also displayed by the decreasing trend in the number of frost ($T_{min} \leq 0$ °C) and freezing days ($T_{min} \leq -10$ °C) | The precipitation regime of the High Tatra mountains did not experience significant changes over the 1966–2006 period, both by amount (−2.6 mm/year) and character of precipitation (extreme and mean). The observed trends are not statistically significant and they explain only a few percent of the variability of particular pluvial characteristics. Seasonally, the precipitation change signals are rather weak (p-value $> 0.05$): winter, spring and summer are generally decreasing, while autumn is on a slightly increase. There are no visible and significant changes in the number of heavy precipitation days ($\geq$30 mm and $\geq$50 mm) | The period of permanent snowpack is shortening in the November-May interval, particularly due to the earlier snowmelt in spring (4.6 days/year). In autumn, snow onset is tending to start earlier at a rate of 0.3 days/year, but this signal is not statistically significant | Żmudzka (2010, 2011) |

| | | | | |
|---|---|---|---|---|
| Bulgarian mountains | An investigation of air temperature changes over a 68 years of observations (1941–2008) at three of the highest mountain peaks in Bulgaria (e.g. Musala 2,937 m, Cherni 2,293 m and Botev 2,378 m) has confirmed the warming trend, accelerated since the late 1970s, at a rate of 0.25–0.30 °C/decade. A dominance of positive anomalies was also observed after 1985. January and summer months are the main contributors to the annual temperature rise. An intensification of zonal circulation over 1941–1950 and 1971–1991 intervals (significant) and a weakening of meridional air flows over 1969–1988 and 2001–2008 intervals (not yet significant) are explaining most of the observed signals at these altitudes in the part of Europe at the border between the temperate and subtropical climatic zones | Humid and dry years alternate on a relatively frequent basis. The periods 1990–1995 and 1998–2002 were generally characterized by high temperatures and low precipitations. Precipitation decrease prevails during summer and autumn seasons while winter and spring show some positive tendencies. A slight decreasing trend was observed in June, while increases were apparent only in September: the statistical significance of these trends is fairly low. The mountain climate seems to be clearly affected by a significant reduction of winter precipitation amounts | The snow cover duration variability over the 1931–2000 period (15 sites) exhibits considerable long-term variability, even if only for few stations; there is evidence of a significant long-term trends (e.g. Samokov and Undola stations) towards an earlier snowmelt. Low elevation sites are not showing an obvious and widespread response to recent warming. Mid-elevation mountain areas (1,000–1,500 m altitude) are experiencing later onsets of snow (in autumn). At Musala Peak (2,937 m) there is a clear reduction in the seasonal frequency of snowfall days, also observed in summer and autumn. Comparatively, rainfall days and air temperature have significantly increased | Koleva and Iotova (1992), Velev (1996), Petkova et al. (2004, 2010), Brown and Petkova (2007), Petkova and Alexandrov (2012), Nojarov (2012), Grunewald et al. (2009) |

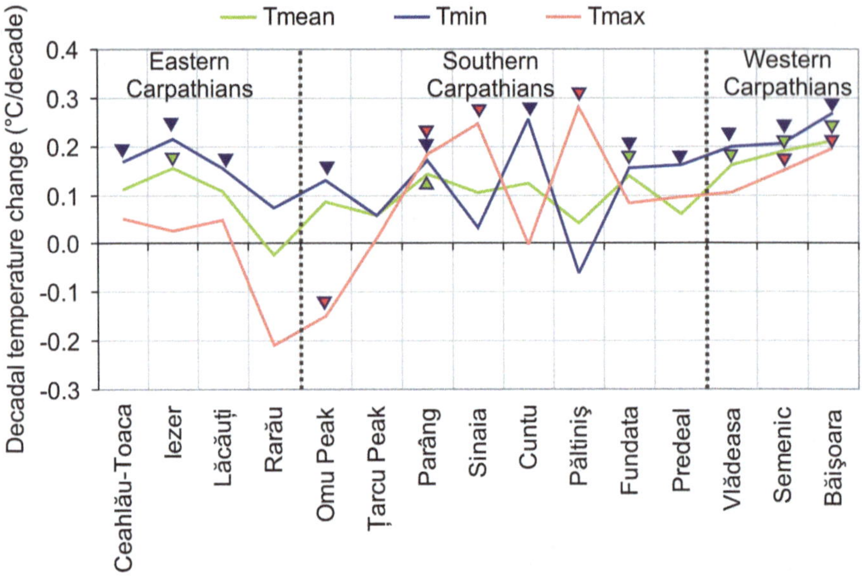


**Fig. 8.1** Decadal temperature change rates (°C/10 years) at some stations in the Romanian Carpathians since 1961. The significant changes ($p < 0.01$) are marked by *triangles*

A closer investigation of daily temperature extremes provides evidence of a greater increase in $T_{min}$ (from 0.14° up to 0.35 °C/decade) than in $T_{max}$ (from 0.19 °C up to 0.27 °C/decade) across the Carpathians. Similar patterns have been also suggested by several works focused on different mountain regions (Karl et al. 1993; Karl et al. 1996; Diaz and Bradely 1997; Beniston 1994; Beniston et al. 1997; Beniston and Rebetez 1996 etc.). Table 8.2 summarizes the annual and seasonal trends statistics and their significance for each individual Carpathian site, according to the Mann-Kendall (MK) trend test, as well as the Theil-Sen slope estimate over 1961–2010. The transition towards a warmer climate seems more rapid in medium and low-to-medium elevation areas of the Southern and Western Carpathians (below 1,600 m). In the Eastern Carpathians (the coldest region), this temperature variability pattern is more blurred.

The seasonal trends in mean temperature are presented in Fig. 8.2. There are exclusively increasing trends in all season except autumn, the most dramatic increases taking place during summer. Autumn is the only stable season with respect to thermal regime.

Warming became obvious and significant from observational records during the 1979–1999 interval in most areas. This early trend maintained its evolution until the end of the study period, when an increase in the frequency of warm and very warm years was noticed all over Carpathians (e.g. 1990, 2000, 2007), but also in other Romanian and European Alpine regions.

The temperature decadal variability in the Romanian Carpathians shows an alternation of cold and warm periods until late 1980s, as well as in early 1990s, when temperature rise became notable at most stations. In high-elevation areas

**Table 8.2** Summary of the annual and seasonal trends in mean air temperature in the Romanian Carpathians

| Station code | Station name | Annual | | DJF | | MAM | | JJA | | SON | |
|---|---|---|---|---|---|---|---|---|---|---|---|
| | | MK score | Theil-Sen slope | MK score | Theil-Sen slope | MK score | Theil-Sen slope | MK score | Theil-Sen slope | MK score | Theil-Sen slope |
| Eastern Carpathians | | | | | | | | | | | |
| 656555 | Ceahlău-Toaca (1,897 m) | **2.11** | 0.014 | 1.32 | 0.016 | 1.09 | 0.015 | **3.65** | 0.038 | −0.67 | −0.008 |
| 737439 | Iezer (1,785 m) | **3.04** | 0.021 | **2.09** | 0.026 | **1.67** | 0.018 | **3.90** | 0.043 | 0.13 | 0.002 |
| 551621 | Lăcăuţi (1,776 m) | **2.69** | 0.019 | **1.66** | 0.022 | **1.77** | 0.023 | **4.55** | 0.043 | 0.12 | 0.000 |
| 719507 | Poiana Stampei (719507) | **3.96** | 0.024 | **1.92** | 0.033 | **2.21** | 0.022 | **4.83** | 0.040 | 0.57 | 0.005 |
| 642540 | Joseni (642540) | **3.36** | 0.024 | **2.01** | 0.047 | **2.38** | 0.029 | **4.33** | 0.039 | 0.03 | 0.002 |
| 541601 | Întorsura Buzăului (707 m) | **3.41** | 0.021 | 1.07 | 0.021 | **1.69** | 0.21 | **5.03** | 0.042 | 0.77 | 0.007 |
| 655522 | Topliţa (687 m) | **3.61** | 0.025 | 1.64 | 0.36 | **2.36** | 0.027 | **4.83** | 0.040 | 0.40 | 0.03 |
| 622544 | Miercurea Ciuc (661 m) | **2.04** | 0.014 | 1.00 | 0.022 | 1.47 | 0.015 | **4.42** | 0.036 | −0.49 | −0.004 |
| 600608 | Târgu Secuiesc (568 m) | **2.99** | 0.020 | 1.25 | 0.022 | **1.76** | 0.020 | **4.68** | 0.041 | 0.12 | 0.001 |
| 542532 | Braşov (534 m) | **2.33** | 0.016 | 0.89 | 0.019 | 1.54 | 0.017 | **4.32** | 0.038 | −0.37 | −0.003 |
| 605537 | Baraolt (508 m) | **3.30** | 0.020 | 1.29 | 0.025 | **2.12** | 0.021 | **4.67** | 0.038 | −0.05 | −0.001 |
| 747356 | Ocna Şugatag (503 m) | **2.81** | 0.020 | 1.41 | 0.026 | **1.91** | 0.020 | **3.81** | 0.037 | 0.27 | 0.002 |
| 758355 | Sighetul Marmaţiei (275 m) | **2.91** | 0.021 | 1.62 | 0.031 | **1.87** | 0.018 | **4.17** | 0.038 | −0.07 | −0.001 |
| 740330 | Baia Mare (224 m) | **2.43** | 0.018 | 1.61 | 0.031 | 1.42 | 0.016 | **3.18** | 0.031 | −0.25 | −0.002 |
| Southern Carpathians | | | | | | | | | | | |
| 527527 | Vf. Omu (2,504 m) | **2.24** | 0.014 | 1.29 | 0.015 | 0.13 | 0.001 | **4.75** | 0.045 | −0.17 | −0.002 |
| 515231 | Ţarcu (2,180 m) | **1.81** | 0.011 | 0.99 | 0.010 | 0.85 | 0.011 | **3.70** | 0.035 | −1.07 | −0.011 |
| 523328 | Parâng (1,548 m) | **2.88** | 0.019 | **2.06** | 0.023 | 1.57 | 0.016 | **4.63** | 0.044 | −0.57 | −0.006 |
| 523530 | Sinaia 1,500 m (1,510 m) | **2.51** | 0.017 | 1.43 | 0.019 | 1.27 | 0.017 | **4.47** | 0.043 | −0.35 | −0.002 |

(continued)

**Table 8.2** (continued)

| Station code | Station name | Annual | | DJF | | MAM | | JJA | | SON | |
|---|---|---|---|---|---|---|---|---|---|---|---|
| | | MK score | Theil-Sen slope | MK score | Theil-Sen slope | MK score | Theil-Sen slope | MK score | Theil-Sen slope | MK score | Theil-Sen slope |
| 518231 | Cuntu (1,456 m) | **2.76** | 0.016 | 1.34 | 0.018 | 1.32 | 0.014 | **4.07** | 0.040 | −0.45 | −0.005 |
| 539357 | Păltiniş (1,453 m) | 1.59 | 0.010 | 1.11 | 0.016 | 1.00 | 0.013 | **4.12** | 0.038 | −1.30 | −0.017 |
| 528518 | Fundata (1,384 m) | **2.98** | 0.020 | **2.26** | 0.031 | 1.61 | 0.021 | **4.15** | 0.042 | −0.20 | −0.001 |
| 530535 | Predeal (1,090 m) | **2.54** | 0.015 | 1.59 | 0.023 | 1.20 | 0.015 | **4.33** | 0.038 | −0.55 | −0.006 |
| 525323 | Petroşani (607 m) | **3.75** | 0.021 | 1.51 | 0.021 | **2.19** | 0.020 | **5.05** | 0.042 | 0.00 | 0.000 |
| 525358 | Voineasa (573 m) | **3.20** | 0.019 | 1.07 | 0.015 | **2.38** | 0.024 | **4.80** | 0.037 | 0.25 | 0.002 |
| 538416 | Boiţa (523 m) | **2.48** | 0.019 | **1.82** | 0.028 | **1.69** | 0.018 | **3.63** | 0.037 | −0.22 | −0.003 |
| 553254 | Deva (240 m) | **3.36** | 0.020 | 1.59 | 0.028 | **2.14** | 0.021 | **3.88** | 0.036 | 0.08 | 0.001 |
| Western Carpathians | | | | | | | | | | | |
| 646247 | Vlădeasa 1,800 m (1836 m) | **3.01** | 0.021 | **2.01** | 0.024 | **2.02** | 0.021 | **4.22** | 0.44 | −0.28 | −0.005 |
| 507158 | Semenic (1,432 m) | **3.08** | 0.021 | **2.19** | 0.028 | **2.01** | 0.023 | **3.97** | 0.042 | −0.72 | −0.009 |
| 634322 | Băişoara (1,360 m) | **3.56** | 0.023 | **2.36** | 0.033 | **2.11** | 0.024 | **4.55** | 0.045 | −0.12 | −0.002 |
| 622303 | Câmpeni (Bistra) (591 m) | **3.31** | 0.21 | 1.62 | 0.029 | **2.02** | 0.020 | **4.58** | 0.038 | −0.23 | −0.002 |
| 502141 | Oraviţa (309 m) | **2.61** | 0.019 | 1.05 | 0.18 | **1.98** | 0.023 | **3.25** | 0.036 | −0.40 | −0.007 |
| 610244 | Ţebea (273 m) | **3.36** | 0.019 | 1.56 | 0.028 | **1.91** | 0.020 | **4.32** | 0.039 | −0.03 | −0.001 |
| 455200 | Bozovici (256 m) | **1.74** | 0.011 | 1.17 | 0.017 | 1.42 | 0.012 | **2.88** | 0.027 | −1.12 | −0.013 |
| 525215 | Caransebeş (241 m) | **2.71** | 0.017 | 1.39 | 0.023 | **2.21** | 0.020 | **3.60** | 0.035 | −0.45 | −0.008 |
| 444127 | Moldova Veche (82 m) | **3.06** | 0.019 | 1.41 | 0.024 | **2.14** | 0.022 | **3.40** | 0.038 | −0.42 | −0.006 |

Statistically significant trends at 10 % level are shown in bold

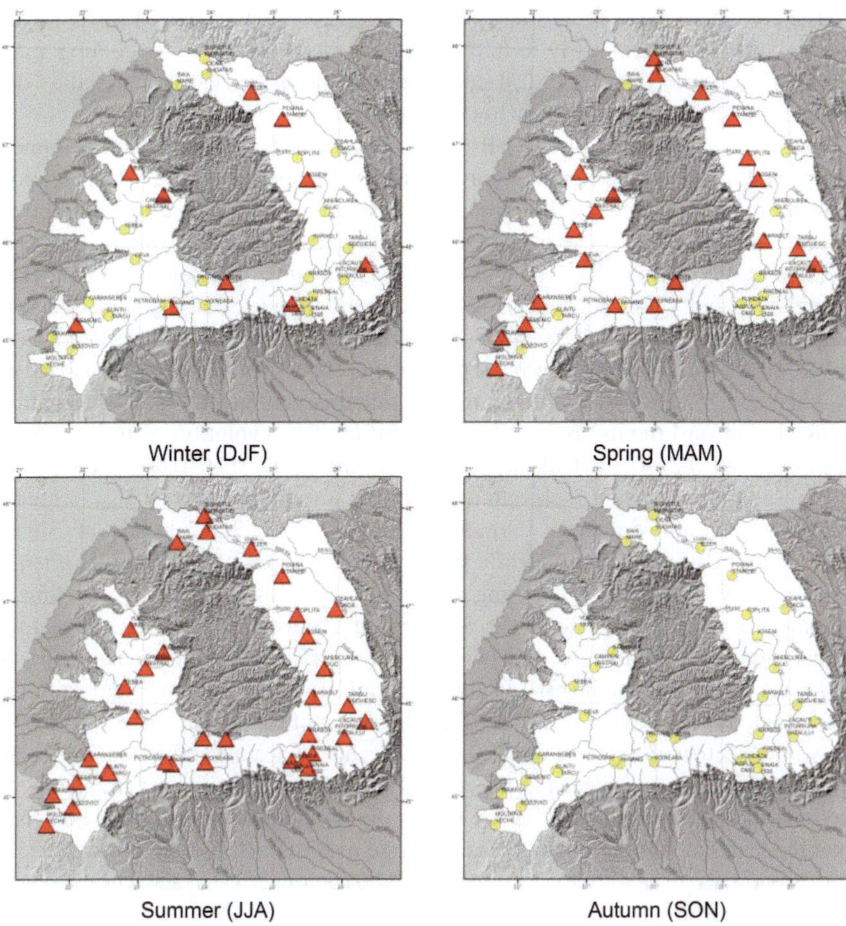


**Fig. 8.2** Seasonal trends in mean air temperature. *Upward red triangles* denote increasing trends, and *light yellow circles* symbolize no trend

(above 2,000 m), the warming process intensified after 1994 but mainly after 1999. The first decade of the twenty-first century (2000–2007) was the warmest since the initiation of meteorological measurements across the Romanian Carpathians. At most sites, this interval was even warmer than the 1990s. These findings are in agreement with the results reported by EEA (2010), WMO (2011) and IPCC (2007), when referred to European and global temperature change until present. Nevertheless, the year 2007 contributed the most to the intensification of warming in the Carpathians, entailing a higher frequency of warm extremes (heat spells, summer days, tropical days and nights) in many regions of Romania, not only in the mountains. More than half of the mountain stations within the 1,100–2,200 m elevation registered historical temperature records in the summer of 2007 (above 25 °C in July).

The cold intervals over the study period, concentrated in the early-to-mid 1960s, most of the 1970s and early-to-mid 1980s. In these decades, the cooling rates ranged between 0.5 and 0.9 °C/decade and the thermal negative anomalies did not exceed 2 °C against the WMO reference average.

Warming has marked almost all *seasons* over the last five decades Summer, winter and spring are generally more subject to temperature rise, while autumn proved rather on a slight cooling trend (yet insignificant). The seasonal variability patterns are consistent those observed at country scale over the same observation period (Busuioc et al. 2010). The highest increase of air temperature is in summer at a rate of about 0.2–0.6 °C/decade, particularly in the Southern and Western Carpathians (e.g. up to 0.59 °C/decade at Sinaia 1,500 m by $T_{max}$, 0.55 °C/decade at Păltiniş by $T_{min}$ – the Southern Carpathians; up to 0.42 °C/decade at Băişoara by $T_{avg}$ and $T_{min}$). The intensity of warming diminishes with height. In high-elevation areas, the upward trends are mostly explained by $T_{min}$ evolution, with increasing rates of about 0.29–0.36 °C/decade at over 2,000 m and 0.36–0.40 °C/decade at 1,700–1,900 m. At medium elevations, warming is more obvious by $T_{max}$ records, particularly in the Southern Carpathians (e.g. 0.55–0.59 °C/decade at Sinaia 1,500 m and Păltiniş stations), while the upward variations of $T_{min}$ are less intense (0.34–0.48 °C/47 years) but covering more extended areas. Summers are tending to become warmer particularly in the Southern and Western Carpathians, due to a higher frequency of intense and persistent tropical and Mediterranean airflows (from the south and southwest). In the Eastern Carpathians summer temperatures are on an increase, mostly due to $T_{min}$ values, more evident in the south (the Curvature sector) at a rate of about 0.40 °C/decade (Lăcăuţi) and less in the north at rates of 0.36–0.38 °C/decade (Ceahlău-Toaca, Iezer). The higher frequency of cold maritime and continental airflows from the north and east of Europe and Asia over the year explain these temperature change patterns in this region. Summers in the Carpathians grew warmer since the mid 1980s below 1,700 m and early 1990s at over 1,800–1,900 m.

Temperature change signals across the Carpathians are not entirely consistent with those observed at country level (Busuioc et al. 2010), warming being more significant in winter than in summer in most regions. Winters tended to be milder in high-elevation areas (above 2,000 m), even if MK estimations suggested no statistical significance. Below 2,000 m, winters became warmer since 1987 (Fig. 8.3). This variability pattern shows great spatial differences across the Carpathians and it was observed both in the $T_{min}$, but mostly in the $T_{max}$ evolution: (i) some areas of 1,300–1,500 m of the Western Carpathians showed more significant increases of $T_{min}$, with confidence at different significance levels (e.g. Semenic 0.27 °C/decade, p-value $< 0.05$; Băişoara 0.29 °C/decade, p-value $< 0.01$), (ii) all the mountain sites surveyed provided significant change rates, significant at the 5 % level, ranging from 0.31 °C/decade in the Western Carpathians (e.g. Semenic and Băişoara stations) to 0.34–0.36 °C/decade in the Southern Carpathians (e.g. Parâng, Sinaia 1,500 m, Păltiniş).

The signals of temperature change are not homogeneous during the extended cold season (November-April). Generally, temperature records illustrate the

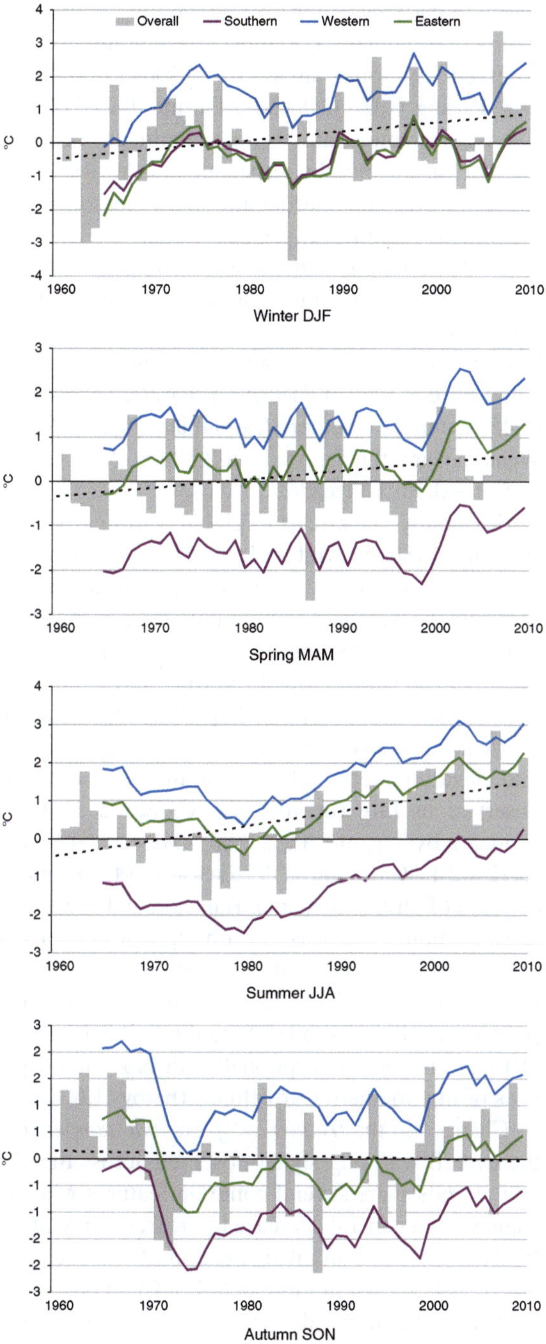


**Fig. 8.3** Regional seasonal mean temperature anomalies (5-year moving averages) in the Romanian Carpathians (1961–2010) relative to 1961–1990. The columns show the general seasonal anomalies over the region

warming patterns better than the cooling ones, even if they are not so strong as in summer or winter. The most outstanding temperature increase (by $T_{max}$) was observed in some areas of 1,300–1,600 m of the Southern Carpathians (e.g. Parâng 0.31 °C/decade, Păltiniş 0.21 °C/decade). Exceptionally, the alpine area of these mountains is on a significant cooling (−0.27 °C/decade) during this interval (in all months, especially in April), as the $T_{max}$ values indicate. Above 2,000 m, warming during the snow season is barely visible by $T_{min}$, and not statistically significant (p-value $< 0.05$).

While autumn is under a slight cooling in several regions of Romania (Busuioc et al. 2010), temperature variations above 1,000 m indicate rather an opposite pattern in this season, still sparsely observed and weaker than in other seasons. Warming in autumn is more noticeable by $T_{max}$ and only in a few areas of 1,300–1,600 m: e.g. 0.59 °C/decade at Parâng (1,585 m) and 0.27 °C/decade at Păltiniş (1,454 m) in the Southern Carpathians; 0.21 °C/decade at Băişoara (1,384 m) in the Western Carpathians. At about 2,500 m (Vf. Omu station), autumns are cooling at a rate of −0.36 °C/decade.

Seasonally, the effects of global warming became visible across the Carpathians since the 1980s in winter (1983–1987) and late 1980s (after 1987) or early 1990s (after 1991) in summer and spring, but more spatially differenced. In autumn, the shifts in the seasonal temperature variability over the period are very spatially differenced, the cooling process being more obvious after 1988, 1995 or exceptionally even earlier after 1968 (only at Parâng station).

*On a monthly basis*, the Carpathian region is warming during more than half of the year, generally from December to August, mostly in mid-elevation and low-to-mid elevation areas (below 1,500 m). Most outstanding temperature increase were observed in July (up to 0.46 °C/decade at Ţarcu by $T_{min}$, 0.51 °C/decade at Vlădeasa by $T_{avg}$ and $T_{min}$, 0.65 °C/decade at Sinaia 1,500 m by $T_{max}$) and January (0.42–0.51 °C/decade at Vlădeasa 1,800 m and Iezer by $T_{min}$, 0.46 to 0.68 °C/decade at Păltiniş, Semenic, Băişoara and Predeal stations). During the peak of the snow accumulation season warming is less intense but still significant, particularly in terms of $T_{min}$ (late winter – February, and early or late spring – March and May). In autumn, the noise is very pronounced and the regional temperature change patterns are rather subtle. A cooling signal is mostly observed in September and October and only exceptionally, in November, particularly at medium elevation sites of the Southern Carpathians (below 1,600 m).

Changes in the *frequency of extreme temperature days* are also consistent with the overall warming signal observed across the Carpathians. The upward temperature trends share several common features with other European mountain ranges, suggesting a general decrease in the frequency of cold extremes (e.g. freezing days, frosty nights) and an intensification of warm extremes (e.g. summer days, tropical days), as is shown in more detail in Chap. 9.

The cold extreme days are under a generalized decrease at most sites of these mountains, mainly after 1990 (or 2000 at some locations). The most significant changes were observed in freezing day frequency ($T_{min}$ below 0 °C) and less in the frosty nights ($T_{min}$ below −10 °C). However, significant reductions of freezing days

($p\text{-value} < 0.05$) were noticed sparsely within the Carpathian region, according to the MK estimations: e.g. 5.40 days/decade at Vf. Omu (2,504 m, the Southern Carpathians), 3.23 days/decade at Băişoara (1,384 m, the Western Carpathians). High-elevations of the Eastern Carpathians (Ceahlău-Toaca, 1,838 m) experience also a decrease in the frequency of these extremes (−2.87 days/decade) but this signal is significant only at the 10 % level of confidence. Less homogeneous changes were observed in the variability of frosty nights ($T_{min}$ below −10 °C), yet statistically insignificant at all sites. Most visible reductions of the number frosty nights were observed in the Southern Carpathians (Vf. Omu, Cuntu, Fundata and Predeal stations), as well as in the Banat Mountains – Western Carpathians (Semenic) and Curvature Carpathians – Eastern Carpathians (Lăcăuţi).

The occurrence of extremely warm days is a valuable indicator of the ongoing warming process across the Romanian Carpathians. In the areas above 800 m, summer days ($T_{max}$ above 25 °C) may be recorded rarely up to 1,800 m, only during the warm years, their peak frequency being a common feature mostly of the mountain areas below 1,500 m and of the foothills below 800 m. The highest frequency of summer days in the areas above 800 m was up to 11 days/year, at Predeal (1,090 m) in the Southern Carpathians.

2007 was a year of thermal record, particularly from May to August, at 12 stations located in the Romanian Carpathians, but also in Romania (Busuioc et al. 2010) and Europe (Annual Bulletin on the Climate of WMO Region VI 2007). The summer of 2007 was particularly warm in response to the occurrence of several intense heat spells, which led to the most severe drought in Romania in the last 60 years and significant crop losses. A high frequency of summer days and even tropical days ($T_{max}$ above 30 °C) and nights ($T_{min}$ above 20 °C), were recorded in the summer of 2007 in many areas below 1,700 m of the Romanian Carpathians. In some mountain areas up to 1,500 m of the Western and Southern Carpathians (Băişoara, Păltiniş, Fundata, Predeal stations), the frequency of tropical nights reached up to two to five cases/month, with a peak in July.

Over the last two decades of the study period (mostly after 1999), several changes in the European-scale circulation were observed, such as a prevalence of the anticyclonic regime, along with a slightly intensification of westerly (of maritime air masses) and southerly airflows (continental air masses from North Africa). Busuioc et al. (2010) emphasized that these changes made it easier to identify a slowly-emerging transition towards milder winters and warmer summers in the Carpathians (particularly evident in the Southern and Western Carpathians). Likewise, milder winters and excessive summers became more frequent in the western and southern plain regions of Romania

The temperature trends over the 50-year period in the Romanian Carpathians are in good agreement with those derived from instrumental records in other European mountain ranges. Temperatures have risen by up to 1 °C in many areas of the Carpathians over this period. This is at a lower rate than in the European Alps during the entire twentieth century (up to 2.0 °C between 1901 and 2000), but a higher rate than the global average temperature change of 0.6 °C/100 years (IPCC 2007, 2013; EEA 2010). Generally, over the same period (1961–2001) temperature oscillations are rather similar, showing a delay in the beginning of warming intensification: earlier in the Swiss Alps (early 1990s) and later in the Romanian Carpathians (late 1990s). Moreover, there is a good correlation with the beginning of the notable temperature rise at global scale (since the 1990s). As in the Swiss Alps (Beniston 2000, 2006), the warming is stronger by $T_{min}$ in most of the seasons, particularly in winter, while $T_{max}$ rise is rather moderate, more evident only in summer (both in the Alps and the Carpathians).

## 8.2 Precipitation

The response of precipitation variation to warming is more difficult to assess, particularly in the regions of complex topography, due to the high spatial and temporal heterogeneity of this variable. Over the last century, weak or non-significant regional changes in precipitation (both amounts and related extremes) have been observed over several mountain ranges of Europe.

The general pattern of annual precipitation variability in the Romanian Carpathians indicates a stable long-term situation, with a very slight increase over 1961–2010 (due to the increasing autumn rain amount), but not significant for most stations. According to the MK statistics, at only 8 out of the 35 weather stations the signal is statistically significant at the 10 % confidence level (two-tail): seven stations show increasing trends (Moldova Veche, Câmpeni, Baişoara, Topliţa, Baia Mare, Ocna Şugatag, Sighetul Marmaţiei) and one station (Vf. Omu) presents a decreasing trend (Table 8.3).

The inter-annual oscillations of precipitation amounts cover an extended variation range and show a great concentration of wet years particularly in the late 1960s and 1970s and sparsely, after 1990 (e.g. 1995, 2005). The drying trend of the Carpathian climate is particularly obvious from observational records since the mid 1980s or early 1990s in most areas, with shifting years covering the 1978–1982 interval.

*Seasonally*, precipitation variability shows considerable spatial differences across the Carpathians and the signals are not homogeneous and consistent as for the temperature regime. A transition towards a drier mountain climate became visible above 1,000 m altitude from 1980 onwards. Generally, winters, springs, summers and even the extended cold season (November-April), grew drier according to both linear regression and Theil-Sen slope estimates, while autumns showed predominant increasing trends. The decline of precipitation amount is comparable in winter, summer and spring, with an average slope of 0.1. However, the Mann-Kendall trend test shows that only in autumn are there statistically-significant increasing trends (apart from 1 to 3 stations showing trends in other seasons, which are reasonable for a 10 % significance level) (Fig. 8.4).

The *monthly* variation of precipitation amount shows increasing trends in March (in Southern and Eastern Carpathians) and in September – in all three mountain regions. Decreasing trends were found in May in the southwestern Carpathians.

**Table 8.3** Summary of trend results in annual and seasonal precipitation in the Carpathians

| Station code | Station name | Annual | | DJF | | MAM | | JJA | | SON | |
|---|---|---|---|---|---|---|---|---|---|---|---|
| | | MK score | Theil-Sen slope | MK score | Theil-Sen slope | MK score | Theil-Sen slope | MK score | Theil-Sen slope | MK score | Theil-Sen slope |
| Eastern Carpathians | | | | | | | | | | | |
| 656555 | Ceahlău-Toaca (1,897 m) | −1.27 | −1.57 | *−3.78* | −1.43 | *−2.38* | −1.21 | 1.39 | 1.24 | 0.60 | 0.28 |
| 737439 | Iezer (1,785 m) | −0.67 | −1.49 | −1.46 | −1.01 | −0.43 | −0.46 | −1.38 | −1.48 | 1.37 | 1.76 |
| 551621 | Lăcăuţi (1,776 m) | −1.47 | −2.98 | −1.26 | −0.39 | −0.49 | −0.19 | −0.80 | −0.94 | −0.37 | −0.22 |
| 719507 | Poiana Stampei (719507) | 0.17 | 0.35 | −0.79 | −0.23 | 0.23 | 0.13 | −0.52 | −0.38 | 0.90 | 0.48 |
| 642540 | Joseni (642540) | 0.63 | 0.72 | −0.06 | −0.02 | −1.53 | −0.57 | 0.55 | 0.34 | **1.90** | 0.81 |
| 541601 | Întorsura Buzăului (707 m) | −0.03 | −0.02 | 0.36 | 0.09 | −0.25 | −0.16 | 0.32 | 0.25 | 1.23 | 0.69 |
| 655522 | Topliţa (687 m) | **2.39** | 2.59 | 0.01 | 0.00 | 1.61 | 0.52 | 0.56 | 0.44 | **2.48** | 1.20 |
| 622544 | Miercurea Ciuc (661 m) | 1.10 | 0.99 | −0.30 | −0.10 | 0.49 | 0.16 | −0.03 | −0.01 | 1.51 | 0.62 |
| 600608 | Târgu Secuiesc (568 m) | 0.85 | 0.90 | 0.17 | 0.05 | 0.73 | 0.28 | −0.01 | −0.01 | **1.77** | 0.63 |
| 542532 | Braşov (534 m) | 0.18 | 0.10 | 0.22 | 0.05 | −0.52 | −0.23 | 0.10 | 0.06 | 1.23 | 0.63 |
| 605537 | Baraolt (508 m) | 1.23 | 1.13 | −0.30 | −0.09 | −0.08 | −0.03 | 0.82 | 0.47 | **2.61** | 1.01 |
| 747356 | Ocna Şugatag (503 m) | **2.43** | 3.53 | 0.26 | 0.08 | 1.00 | 0.46 | 0.75 | 0.57 | **2.63** | 1.72 |
| 758355 | Sighetul Marmaţiei (275 m) | **2.04** | 3.26 | 0.11 | 0.07 | **1.65** | 1.00 | 0.89 | 0.98 | **2.21** | 1.55 |
| 740330 | Baia Mare (224 m) | **2.90** | 4.41 | 1.30 | 1.05 | 1.52 | 1.10 | 0.50 | 0.47 | **2.93** | 2.04 |
| Southern Carpathians | | | | | | | | | | | |
| 527527 | Vf. Omu (2,504 m) | **−1.87** | −4.96 | −1.59 | −0.90 | −0.31 | −0.16 | **−1.84** | −2.11 | −0.07 | −0.08 |
| 515231 | Ţarcu (2,180 m) | −0.89 | −2.51 | 1.07 | 0.97 | −0.84 | −0.65 | **−1.81** | −2.41 | 1.09 | 0.76 |
| 523328 | Parâng (1,548 m) | −0.85 | −1.64 | −0.18 | −0.14 | −0.12 | −0.07 | −0.82 | −0.93 | 0.38 | 0.36 |
| 523530 | Sinaia 1,500 m (1,510 m) | −1.02 | 2.06 | −0.27 | −0.20 | −1.57 | −1.29 | −0.04 | −0.04 | 1.49 | 1.46 |
| 518231 | Cuntu (1,456 m) | −1.47 | −5.27 | −0.03 | −0.02 | −1.10 | −1.00 | *−1.87* | −3.01 | 0.05 | 0.11 |

(continued)

**Table 8.3** (continued)

| Station code | Station name | Annual | | DJF | | MAM | | JJA | | SON | |
|---|---|---|---|---|---|---|---|---|---|---|---|
| | | MK score | Theil-Sen slope | MK score | Theil-Sen slope | MK score | Theil-Sen slope | MK score | Theil-Sen slope | MK score | Theil-Sen slope |
| 539357 | Păltiniş (1,453 m) | 1.02 | 1.68 | 0.54 | 0.18 | 0.67 | 0.61 | 0.77 | 0.89 | 1.33 | 0.97 |
| 528518 | Fundata (1,384 m) | 0.70 | 1.06 | −0.30 | −0.15 | 0.03 | 0.06 | 0.94 | 0.99 | 1.12 | 0.84 |
| 530535 | Predeal (1,090 m) | 0.03 | 0.11 | −0.40 | −0.21 | −0.35 | −0.25 | 0.43 | 0.51 | 0.93 | 0.64 |
| 525323 | Petroşani (607 m) | −1.62 | −2.47 | −0.83 | −0.35 | −1.00 | −0.51 | −0.42 | −0.41 | 0.07 | 0.09 |
| 525358 | Voineasa (573 m) | −1.52 | −2.41 | −0.52 | −0.31 | −1.47 | −0.85 | 0.35 | 0.28 | −0.23 | −0.18 |
| 538416 | Boiţa (523 m) | 0.33 | 0.44 | −0.38 | −0.17 | −1.00 | −0.50 | 1.19 | 1.07 | 1.21 | 0.83 |
| 553254 | Deva (240 m) | 1.19 | 1.30 | 0.00 | 0.00 | 1.01 | 0.39 | 0.80 | 0.52 | 0.97 | 0.41 |
| Western Carpathians | | | | | | | | | | | |
| 646247 | Vlădeasa 1,800 m (1836 m) | −0.85 | −1.43 | −0.99 | −0.75 | −0.38 | −0.22 | −0.40 | −0.53 | 0.79 | 0.66 |
| 507158 | Semenic (1,432 m) | −0.84 | −2.26 | −0.72 | −0.55 | −1.12 | −1.02 | −1.00 | −1.54 | 0.72 | 0.81 |
| 634322 | Băişoara (1,360 m) | **1.84** | 2.69 | −0.84 | −0.31 | 0.38 | 0.19 | 1.46 | 1.29 | **1.83** | 1.15 |
| 622303 | Câmpeni (Bistra) (591 m) | **2.14** | 3.21 | 1.04 | 0.63 | **1.82** | 1.07 | 0.69 | 0.60 | **1.88** | 1.14 |
| 502141 | Oraviţa (309 m) | 1.12 | 1.84 | 0.62 | 0.44 | 0.44 | 0.36 | 0.18 | 0.30 | **1.84** | 1.19 |
| 610244 | Ţebea (273 m) | 1.44 | 1.77 | −0.62 | −0.30 | 0.55 | 0.29 | 0.28 | 0.20 | **1.99** | 1.13 |
| 455200 | Bozovici (256 m) | 0.25 | 0.23 | 0.00 | 0.01 | −0.90 | −0.40 | −0.08 | −0.07 | 1.15 | 0.57 |
| 525215 | Caransebeş (241 m) | 0.23 | 0.26 | 0.57 | 0.30 | 0.23 | 0.12 | −0.52 | −0.46 | 1.30 | 0.84 |
| 444127 | Moldova Veche (82 m) | **1.69** | 2.08 | 0.59 | 0.29 | −0.17 | −0.06 | 0.18 | 0.10 | **2.06** | 1.22 |

Statistically significant trends at 10 % level are shown in bold

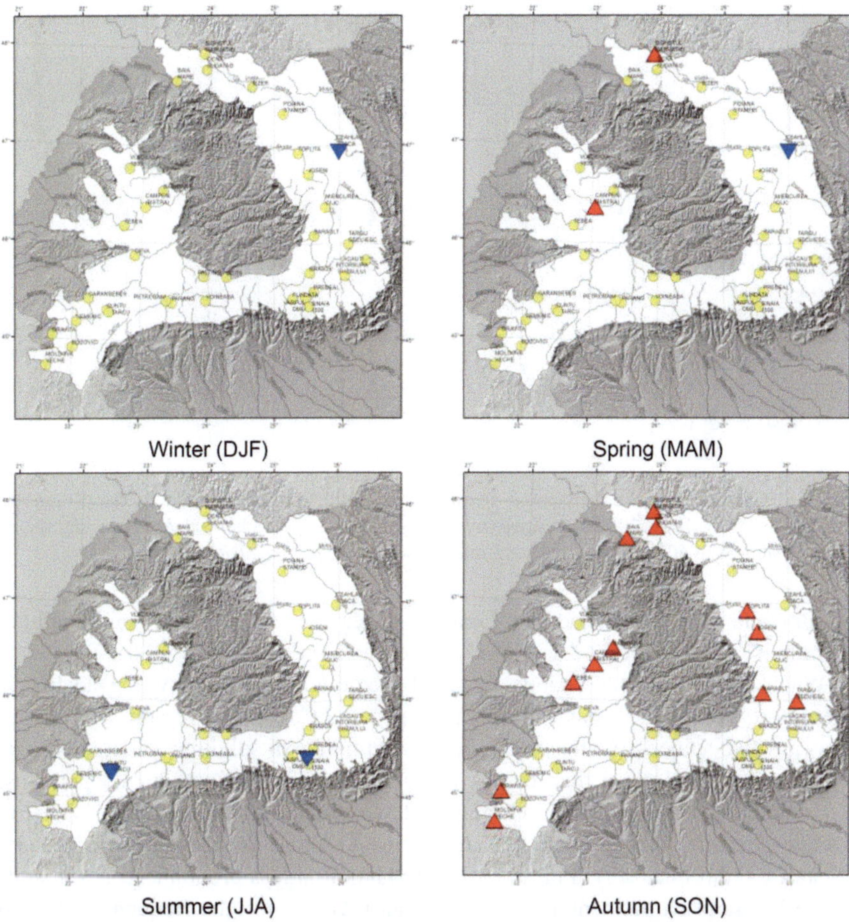


**Fig. 8.4** Trends in seasonal precipitation. *Downward blue triangles* signify decreasing trends, *upward red triangles* denote increasing trends, and *light yellow circles* symbolize no trend

To better outline the response of precipitation regime to the observed climate warming, the year-to-year variability of *solid* (snowfalls, snow showers, sleet and sleet showers, hail) and *liquid precipitation frequency* (e.g. rainfalls, rain showers, drizzle) was investigated. There is a clear connection between the warming intensification since the mid 1980s and early 1990s and the generalized reduction of solid and liquid falls frequency (statistically significant at most of the locations) (Fig. 8.5). The occurrence of liquid falls tend to decline more at high-elevation sites (above 1,700 m) and less at lower elevations (below 1,600 m). Significant changes in liquid fall frequency were detected in the forest areas of the Southern Carpathians (e.g. Cuntu 4.4 days/decade; Sinaia 1,500 m 4.3 days/decade; Predeal 2.8 days/decade) and only sparsely, in their alpine belt (Vf. Omu, 6.5 days/decade). Regionally, the greatest decreases are specific to some high-elevations (7.9 days-decade at

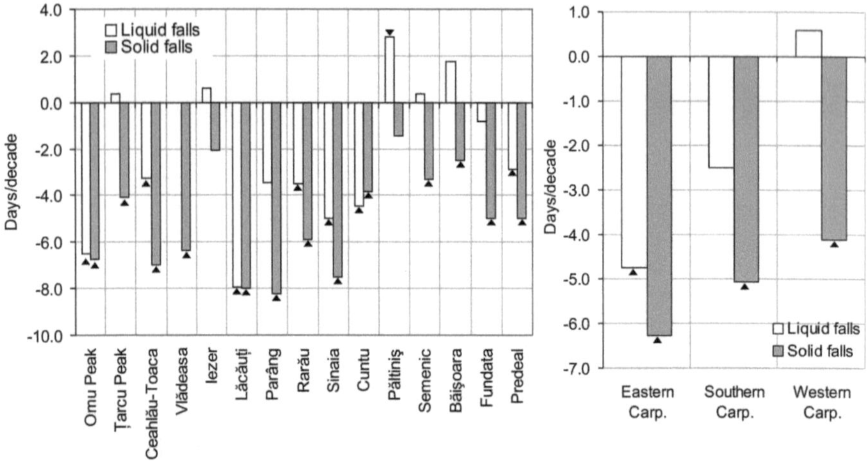


**Fig. 8.5** Change rates in the annual frequency of solid and liquid falls in the Romanian Carpathians over 1961–2007: by sites (*left*); by Carpathian branches (*right*). The significant cases (confident to at least 5 % level) are marked with a *small black triangle*

Lăcăuți and 3.1 days/decade at Ceahlău-Toaca 3.1 days/decade) and medium elevations (2.9 days/decade at Rarău) of the Eastern Carpathians.

When referring to the changes in solid precipitation frequency, the decline is even more pronounced and widespread across the Carpathian region. Mid elevations are also subject to outstanding downward trends of solid hydrometeor occurrences at a rate of more than 6.0 days/decade (e.g. in the Southern Carpathians: Vf. Omu 6.7 days/decade, Parâng 8.2 days/decade and Sinaia 1,500 m 6.4 days/decade; in the Eastern Carpathians: Ceahlău-Toaca 6.55 days/decade; Lăcăuți 8.0 days/decade) (Fig. 8.5). The reduction is less pronounced at the sites of the Western Carpathians (3–5 days/decade). Generally, the frequency of solid falls tends to decrease more in the Eastern and Southern Carpathians than in the Western Carpathians, where on the contrary, liquid falls are on a slight increase (yet insignificant over the study period).

There is a progressive transition from a wetter climate to a drier one in many areas of the Romanian Carpathians, as suggested by a slight expansion of dry spell intervals (up to 2 days/decade), a decline of wet days frequency (up to 11 days/decade) and a shrinking wet spell intervals (0.5 days/decade). These findings are particularly characteristic for the Southern Carpathians and they are significant mostly in summer.

The probability of wet day occurrences is on the decrease at most Carpathian sites, while the average precipitation intensity is slightly and sparsely increasing (more rainfall showers in some areas). In the mountain areas above 1,700–1,800 m and in those below 1,500 m, higher frequency of showers tends to slightly compensate for the decrease in the number of rainfall days. This signal was observed especially in autumn, summer and over less extended areas, in spring. These patterns explain the small precipitation changes at some sites for which the MK test did not provided evidence of significant decrease at annual scale. The annual frequency of rain showers is on a slight increase in some areas at rates of 5.1–9.7 days/decade at high-elevations and 3.9–10.4 days/decade at medium and low-to-medium elevations.

There are no significant oscillations in the extreme precipitation amounts (greatest 1-day, 2-day and 3-day precipitation). The areas of significant change are quite scattered across the Carpathians, most significant changes suggesting a widespread decrease at some locations of the Southern Carpathians. The decrease is significant in the alpine area (p-value $< 0.001$) and particularly stronger over 3-day heavy rainfall spans (7–11 mm/decade). The frequency of heavy precipitation days (more than 10 mm/day) shows also a generalized decrease and this pattern is also more obvious in the Southern Carpathians (above 2,000 m and below 1,600 m), both in summer and winter, at a rate of 1.7–5.9 days/decade. There are no significant changes in the frequency of days with more than 30 and 50 mm over the Romanian Carpathians.

The precipitation change signals observed from instrumental records at selected weather sites are consistent with the results of regional projections under current climate for other European mountain range regarding the decrease in summer precipitation amount.

## 8.3 Wind

The near-surface airflow (at 10 m above the ground) is an important natural forcing factor of the observed climatic oscillations in the Carpathians. Wind is playing a key role in driving the weather conditions in these mountains, shaping the temperature and precipitation regime (both liquid and solid falls) and influencing the evapotranspiration. Changes in mean wind speed at high-elevations are considered relevant attributes of the observed changes in large-scale circulation over Europe. Trends in wind speed were investigated and tested over 1961–2010.

Significant decreases of *annual wind speed* were observed in most areas of the Romanian Carpathians (p-value $< 0.05$). Regionally, the Eastern Carpathians experienced the most significant weakening of wind speed. The observed decreases in wind speed cover the entire elevation range (Fig. 8.6).

The signals derived from the records at high-elevation sites (above 1,700–1,800 m), are reliable for linking local changes in wind speed to changes in large-scale circulation. There is a fairy monotonic increase of wind speed with elevation. A point wise comparison between the annual wind speed changes at some mountain stations (Fig. 8.7) suggests that the near-surface wind speed tends to decline more rapidly at high-elevation sites (above 1,700 m), particularly in the Eastern and Southern Carpathians (e.g. Lăcăuți, Ceahlău, Vf. Omu, Ţarcu). The decrease is also obvious at some mid and high-elevation sites of the Western Carpathians, located between 1,400 and 1,800 m (e.g. Vlădeasa 1,800 m and Semenic stations). These areas are usually located in open conditions, on or near some high-altitude plateaus and mountain peaks. The changes of wind speed at site scale are consistent with the

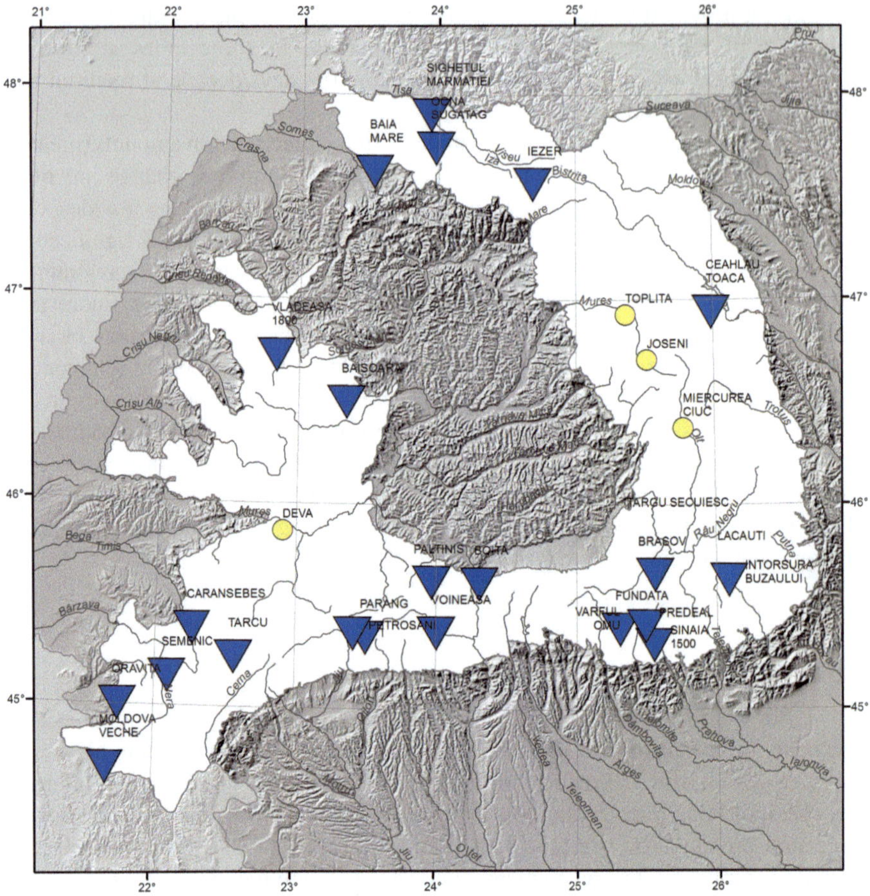


**Fig. 8.6** Annual trends in mean wind speed in the Romanian Carpathians. Significant decreasing trends are marked with *downward blue triangles*

estimated changes in large-scale circulation, towards the prevalence of anticyclone regimes (Dima 2012).

These results are also in total agreement with those obtained by McVicar and Roderick (2010), who determined similar wind speed change patterns based on the observational records (1960–2006), provided by 119 sites located in two different mountain regions (e.g. Swiss Alps, between 203 and 3,580 m and the Loess Plateau of China, between 53 and 3,629 m). The authors emphasized that the decline is more rapid at higher elevations than at lower elevations. There is also a good agreement with the results reported by Busuioc et al. (2010) at national scale. They attributed the wind speed declines to a stronger zonal circulation and a weaker cyclonic activity, particularly in the last decades of the 1961–2010 period (during all seasons, particularly in winter and spring).

*Seasonally*, there is also a significant and widespread weakening of near-surface airflow dynamics across the Carpathians in all seasons (Fig. 8.8). Overall, the trend

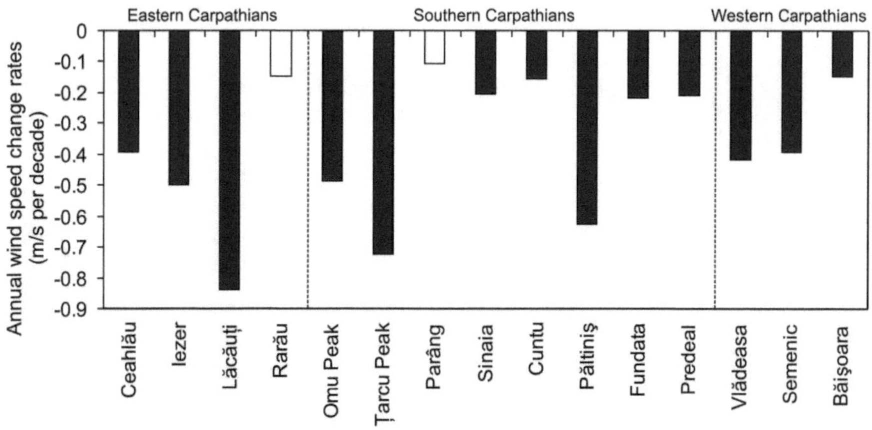


**Fig. 8.7** Decadal change rates of annual wind speed at individual mountain sites (above 800 m) over 1961–2010. The *white columns* depict sites where the MK test did not return significant changes

results fit well within the observed country-wide decrease in average wind speed (Birsan et al. 2013). The results are in agreement with the vast majority of recent studies on wind speed trends, which report an overall tendency towards terrestrial stilling (McVicar et al. 2008). They also point out the importance of taking into account the long-term variability of the seasonal mean wind speed, when assessing long-term changes in wind-related variables – like evaporation or climate extremes indices.

An obvious increase in the frequency of atmospheric calm at high-elevation areas represents another change in wind regime, explained by the strengthening of anticyclonic activity over the last decades. These results are in total agreement with those provided by Busuioc et al. (2010) at country scale, which associated this signal to a temperature-increasing trend at the 850 hPa level. The increase of calm frequency is most obvious in winter (the most dynamic season) at all four sites, at a rate of 1.5–3.5 % per decade. These findings are even more relevant if considering the changes observed at 2,500 m (Vf. Omu, the highest weather site of the Romanian Carpathians). Changes in airflows regime at this alpine site may offer valuable indications of large-scale circulation variability, which may explain the frequency and magnitude of the climate anomalies recorded in the Carpathians over the last decades.

## 8.4 Snowpack

Among the terrestrial components of the cryosphere, snowpack has the largest extent and it is perceived as a relevant indicator of change in the climate system, as it proved to respond sensitively to recent temperature and precipitation oscillations (Global outlook for ice and snow 2007). Air temperature and precipitation are the most responsible for changes in snow coverage and accumulated snow

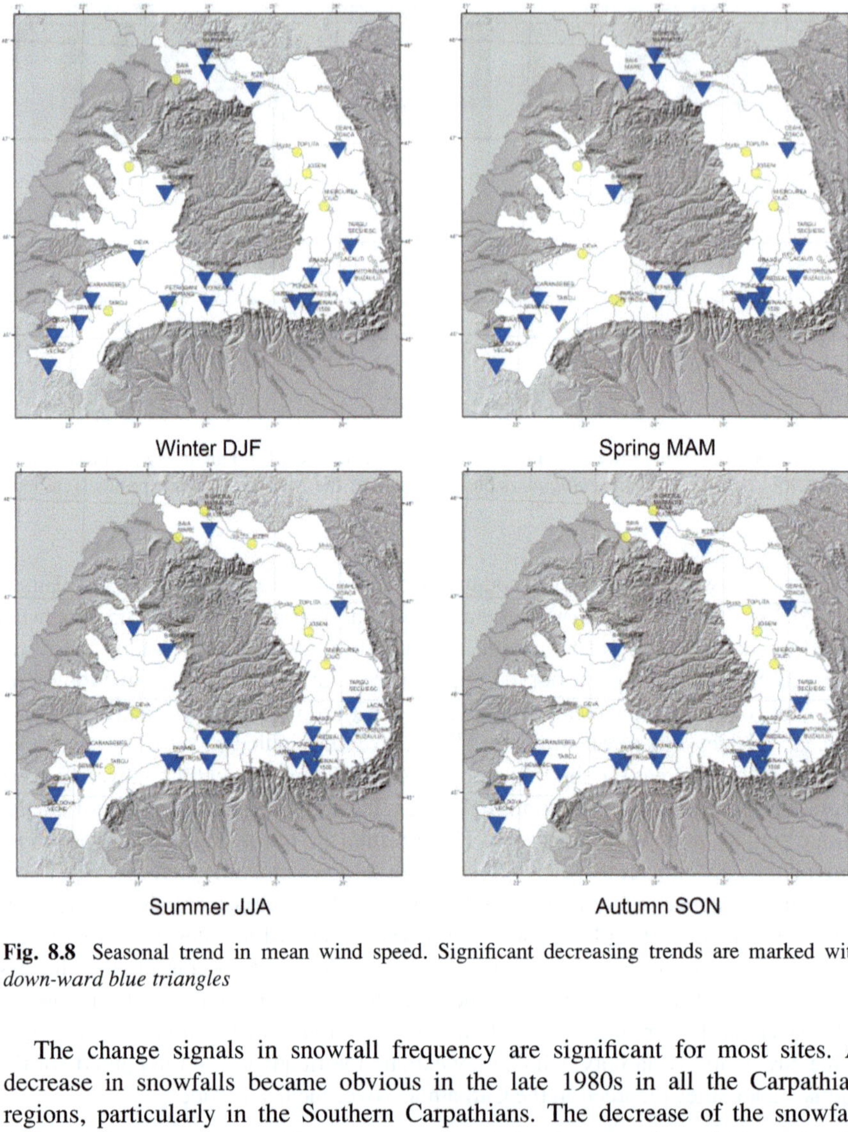


**Fig. 8.8** Seasonal trend in mean wind speed. Significant decreasing trends are marked with *down-ward blue triangles*

The change signals in snowfall frequency are significant for most sites. A decrease in snowfalls became obvious in the late 1980s in all the Carpathian regions, particularly in the Southern Carpathians. The decrease of the snowfall days number was unfavorable to snow accumulation in areas above 800 m, already affected by warming and drying. There was only one winter with positive anomalies of snowfall days (1995/1996), in the last 20 years of the period (up to 10 %).

The Mann-Kendall trend test revealed downward trends in the number of days with snow coverage (Fig. 8.9) at almost one third of the locations; mean snow depth shows similar patterns both annually and for the winter season, with few exceptions (Fig. 8.10).

The most visible change concerns the number of snowfall days – which is decreasing at more than two thirds of the stations (Fig. 8.11), pointing an

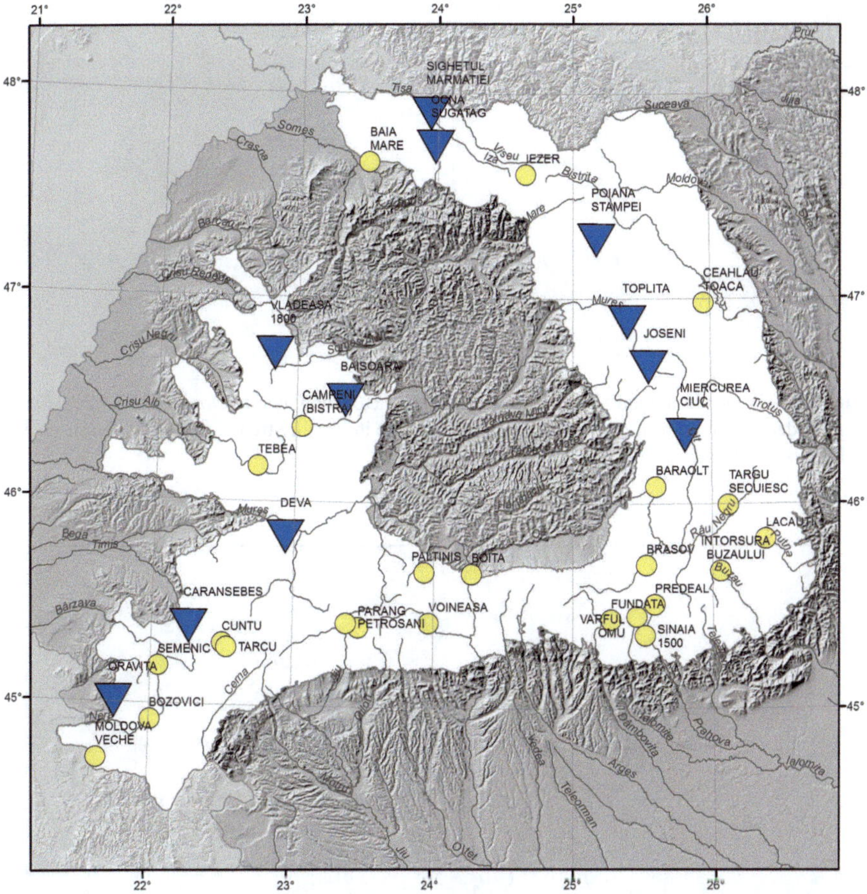


**Fig. 8.9** Trends in the number of days with snow cover during winter (DJF). Significant decreasing trends are marked with *downward blue triangles*

acceleration of the water cycle (Huntington 2006): increased winter temperatures may result in increasing precipitation following the Clausius–Clapeyron relationship, depending on the slope of the snowfall-temperature relationship (Davis et al. 1999). Most of the locations with decreasing trends in snowfall days present no decrease in snow depth, indicating that the same amount of snow tends to fall within a considerably shorter interval. Increases in snowfall intensity could be explained by an increased moisture-holding capacity of the warmer atmosphere. Another explanation for the rather unaltered mean snow depth despite less snowfall days could be that mainly the light snowfall events are affected by transition to rain, while the heavy snowfall events remain unaffected by the transition to rain (Birsan and Dumitrescu 2014).

*Onset and melting dates* of snowpack are important climate and ecological markers (Groisman et al. 1994) and their evolution is visibly linked to the recent

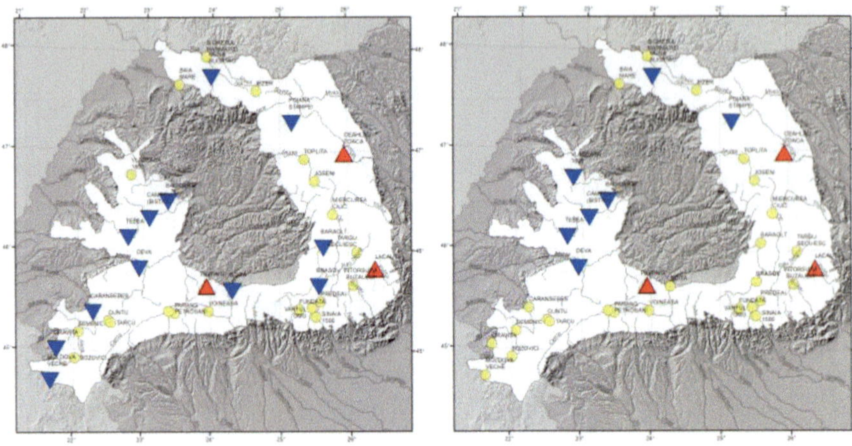

**Fig. 8.10** Trends in annual (*left*) and winter (*right*) mean snow depth. *Downward blue triangles* signify decreasing trends, *upward red triangles* denote increasing trends, and *light yellow circles* symbolize no trend

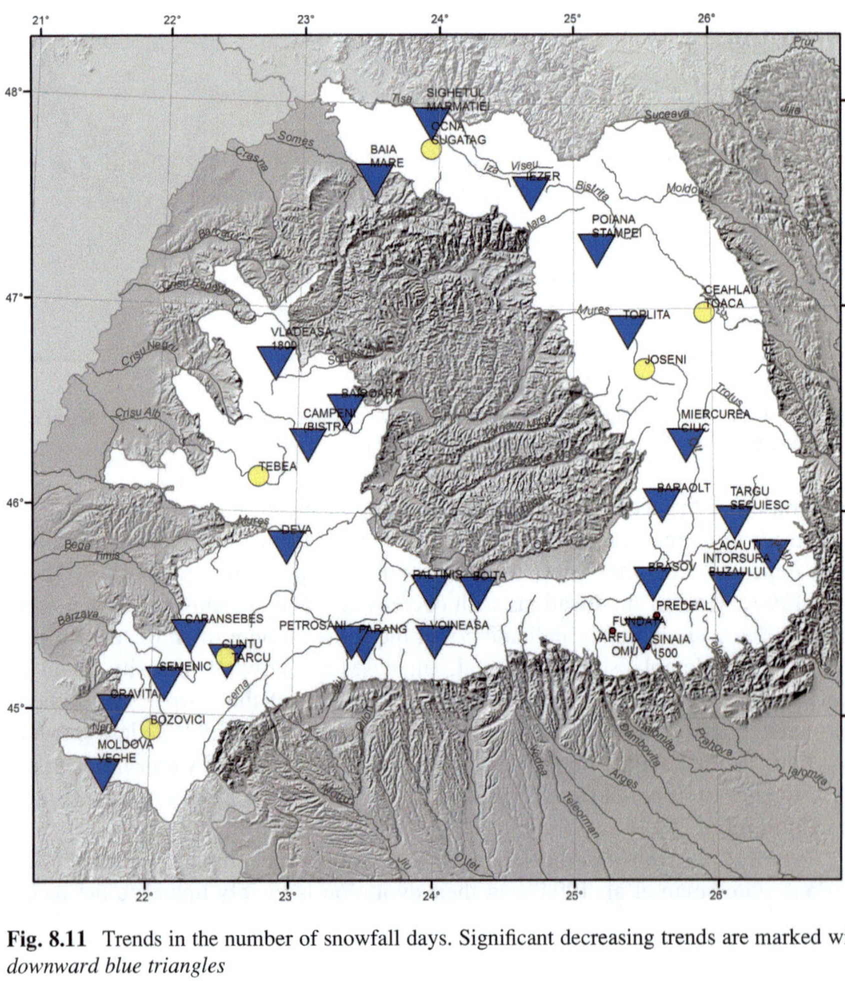


**Fig. 8.11** Trends in the number of snowfall days. Significant decreasing trends are marked with *downward blue triangles*

temperature and precipitation variations over 1961–2003. In autumn, snowpack tended to set in earlier than the average date only at the alpine sites of the Southern Carpathians (e.g. Vf. Omu 6 days/decade and Țarcu 10 days/decade), while at the rest of the stations below 1,800 m (e.g. Lăcăuți, Rarău, Sinaia 1,500 m, Fundata and Predeal), it tended to form later, about 4–5 days/decade. These autumn variability patterns became obvious at the end of 1980s and early 1990s. In spring, there is a generalized trend towards earlier snowmelts all over Carpathians, but stronger in the mid-elevation areas (below 1,600 m) of the Southern and Western Carpathians. In these areas, snowmelt onset tended to occur earlier by 2–8 days/decade. There were no consistent variations of snowmelt dates from the late 1960s to late 1980s. Most changes occurred during the 1990s, which marked the beginning of a visible snow cover duration shortening in most areas. Earlier snowmelt in spring explains the reduction of snow cover duration in the Romanian Carpathians more than the later snow onset in autumn (Micu 2009). This signal is in agreement with recent findings of increasing trends in spring streamflow in natural river basins in the Romanian Carpathians (Birsan et al. 2012, 2014).

A detailed decadal investigation of snow cover duration (SCD) was carried out by Micu (2009). The research was based on the relative deviations of the annual SCD relative to the long-term average (1961–2003), suggesting large inter-decadal oscillations. The positive and negative deviations of the annual snow pack duration are rather scattered between 1961 and 2003 and usually do not exceed the ±40 % threshold. As a general picture, the amplitudes of the highest anomalies in the negative domain (e.g. −44 % at Fundata in the winter 1962/1963) are fairly comparable to those in the positive range (e.g. 44 % at Parâng in the winter 1974/1975).

After 1990, the decline of snow cover duration became more obvious. At most mid-elevation sites, particularly located in the Southern Carpathians, the lack of snow was quite evident. The negative anomalies dominated the last decades (e.g. 1999/2000 and 2000/2001) and the SCD recorded less than 160 days/year at 7 out of the 15 weather sites. This variability pattern was maintained even after 2003 and culminated in the winter 2006/2007. This winter was the mildest ever recorded in most regions of the country, but also in most of European mountain ranges, entailing great loss to winter tourism operators due to the unusual lack of snow (Bogdan et al. 2007). The Carpathians were most affected at medium elevation levels (below 1,600 m), where most ski slopes are located.

Snow depth is highly variable with height across the Carpathians, showing large differences between the sites, in terms of accumulated snow amounts and their spatial distribution. The network of snow measurements is rather sparse in these mountains. For these reasons, the observed trends in snow depth variability are considered representative only at local scale. Despite these limitations, snow depth and its snow-related extremes are considered relevant indicators of a mountain climate that is changing.

The change signals of snow depth have been derived from the variability of the greatest seasonal snow depth and the accumulated seasonal snow amounts, over an extended winter season (October-May), when snow avalanches are the most frequent (the average snow avalanche season). A slight intensification of snow extremes, outlined by the upward trends of maximum seasonal snow depths, was only sparsely observed across the Carpathians (see Chap. 9). The strongest signals of maximum snow depth frequency and magnitude were recorded over December–March and especially, March–May intervals, when the North Atlantic and Mediterranean cyclonic activity is usually intense. The winter-to-winter variation of snow accumulation, based on seasonal accumulated snow amounts, support these findings.

Increased snowy conditions across the Romanian Carpathians were characteristic of the early 1960s (at 1,700–1,900 m), mid 1980s (below 1,600 m) and late 1990s (in some areas below 1,800 m). Snow-scarce winters were more frequently recorded in the mid-to-late 1960s, early 1970s, late 1980s, and early 1990s and after 2000. The 1990–2000 decade was outstanding is terms of snow extremes, when at least five mountain stations simultaneously recorded both the largest (late 1990s) and the lowest maximum snow depths and accumulated snow amounts (early 1990s). The changes in snow accumulation above 1,000 m were investigated in terms of seasonal share of snow accumulation over three winter sub-periods: early winter period (October to December), core winter period (December to March) and late winter period (March to May). This approach was proposed by Schöner et al. (2009). The duration of snow accumulation season in the Romanian Carpathians is tending to shorten. A shift towards less snow accumulation is visible in the second half of the winter season (March to May), while a shift towards more snow deposition was observed only during its first half (October to December). There is a generalized decline in the early winter contribution to the overall seasonal snow accumulation, but less pronounced than in the core winter interval. Most changes were attributed to core and late winter intervals. This shows that mountain areas prone to long-lasting snow accumulation seasons are generally at the mid-to-high elevations (1,700–1,900 m) of the Eastern Carpathians (e.g. Ceahlău-Toaca 18.6–18.7 % per decade; Lăcăuți 6.9–7.3 % per decade) and mid-elevation areas (1,400–1,500 m) of the Southern Carpathians (e.g. Păltiniş 5.9–7.1 % per decade). Nevertheless, some mid-elevations are also marked by a visible and significant decline of snow accumulation in spring, like Rarău and Băişoara. There is a generalized reduction of the snow accumulation season at mountain stations located between 1,300 and 1,900 m altitude. This signal is the response of a lower contribution of early and late winter intervals to the overall snow accumulation, since the 1980s, when the change rates varied from 1 % per decade to 20 % per decade. Most changes became obvious since the early and late 1970s (e.g. 1971, 1978) or early 1990s in the Eastern Carpathians, and early-to-mid 1980s (e.g. 1980, 1984–1985, 1986) in the Southern and Western Carpathians. These trends are linked to the changes in snowfall regime and to the slight intensification of snow extremes in some areas (e.g. more frequent heavy snowfalls).

There is a good agreement with the findings of Bojariu and Dinu (2007), who investigated the snowpack variability in the Romanian Carpathians over 1961–2000, based on the variations of the accumulated snow amounts between November and March. Accordingly, the snow accumulation is on a continuous decrease in January and March, mostly in the northwestern and central parts of the country, but also in the Western Carpathians (e.g. Vlădeasa 1,800 m and Băişoara stations) and northeastern part of the Eastern Carpathians (e.g. Rarău station). However, in most of the ski areas

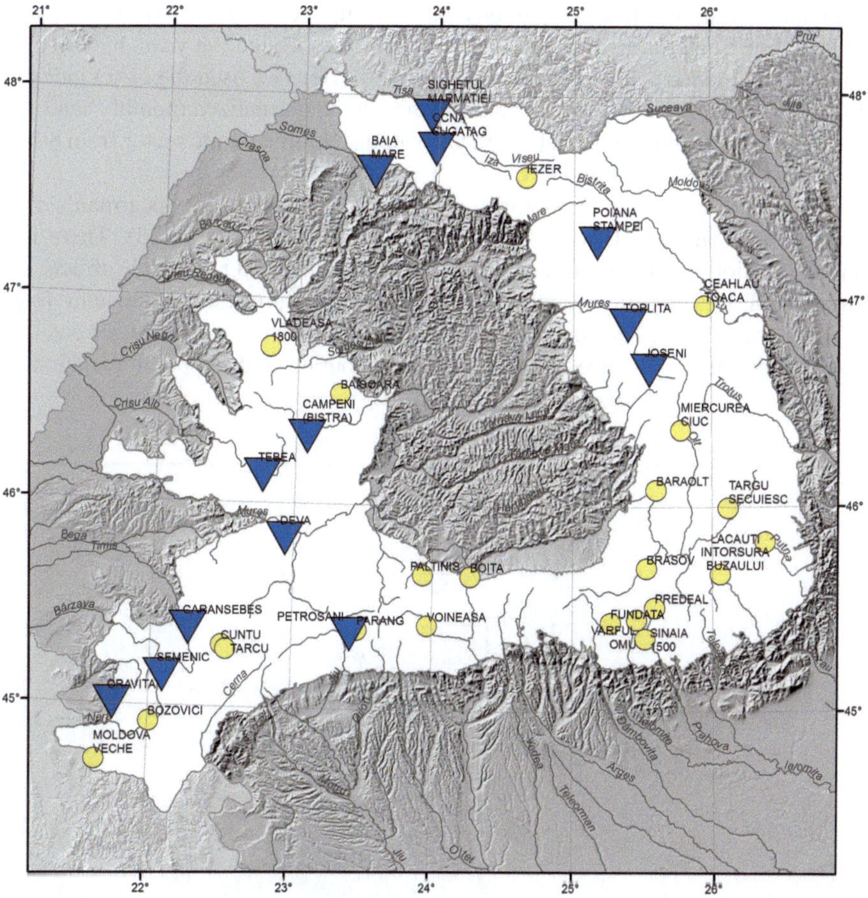


**Fig. 8.12** Trends in continuous snowpack duration (November–April). Significant decreasing trends are marked with *downward blue triangles*

of the Southern and Eastern Carpathians, the seasonal snow depth oscillations were rather constant, with no indication of a significant change. Over sparse mountain areas of the Eastern (Lăcăuţi) and Southern Carpathians (Păltiniş), an upward trend was observed during the entire cold season.

The continuous snowpack duration (maximum number of consecutive days with snow cover from November to April) presents decreasing trends at more than one third of the stations (Fig. 8.12).

## 8.5 Influence of Large-Scale Circulation

The North Atlantic Oscillation (NAO) has been shown to influence the winter precipitation in Europe. Beniston (1997) found the NAO influencing the timing and amount of snow in the Swiss Alps. Bednorz (2002) showed that the NAO exhibited strong negative correlations with the snow cover duration over western Poland. For Bulgaria, Brown and Petkova (2007) associated the years having high snow accumulation with a negative NAO pattern. Here we used the NAO index of Li and Wang (2003), defined as the difference in the normalized monthly sea level pressure (SLP) regionally zonal-averaged over the North Atlantic sector from 80°W to 30°E between 35°N and 65°N.

We found significant monotonic correlation (by means of Spearman's rho) between the winter DJF NAO and the mean snow depth (Fig. 8.13). There is a strong relationship between snow variability and the NAO – which affects the strength of westerly flow and weather patterns in Europe during winter (e.g., Wanner et al. 2001). The advective processes exerted by the large-scale circulation have a dominant influence on the spatial distribution and temporal variation of

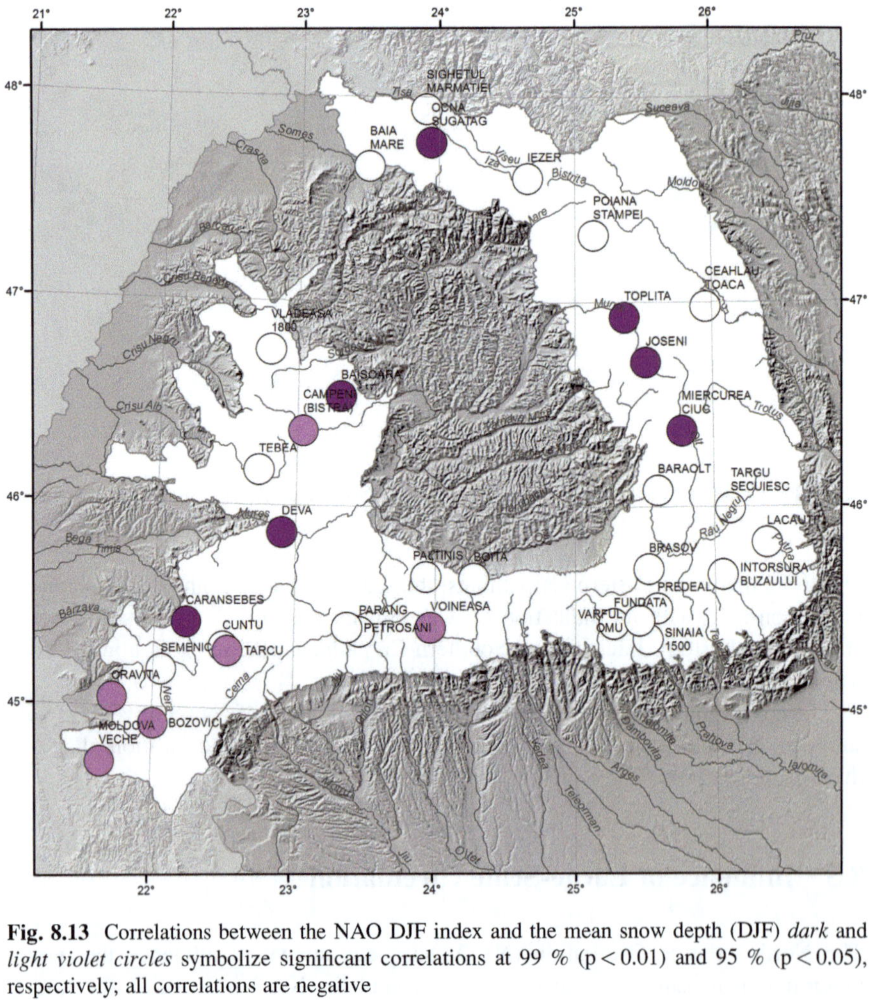


**Fig. 8.13** Correlations between the NAO DJF index and the mean snow depth (DJF) *dark* and *light violet circles* symbolize significant correlations at 99 % ($p < 0.01$) and 95 % ($p < 0.05$), respectively; all correlations are negative

European climate during winter (Küttel et al. 2011). In the Romanian area, the positive thermal anomalies and the negative precipitation anomalies are associated with a high NAO index (Bojariu and Paliu 2001). Negative correlations between the DJF NAO index and the number of days with snow coverage have also been found. The NAO index correlates best with the number of snowfall days and with the precipitation amount (for more than two thirds of the stations).

The climatic variability in the winter season seems to be driven by the large scale circulation over the North Atlantic, the NAO signal being strong enough to (partly) overcome the effect of orography.

## References

Annual Bulletin on the Climate in WMO Region VI – Europe and Middle East (2007). http://www.dwd.de/rcc-cm

Armstrong R, Brodzik MJ (2001) Recent northern hemisphere snow extent: a comparison of data derived from visible and microwave sensors. Geophys Res Lett 28(19):3673–3676

Bárdossy A, Anagnostopoulou C, Cacciamani C, Caspary H, Frei C, Goodess C, Haylock M, Hundecha Y, Maheras P, Michna P, Pavan V, Plaut G, Schmidli J, Schuepbach E, Simonnet E, Tolika K, Tomozeiu R (2003) Trends in extreme daily precipitation and temperature across Europe in the 2nd half of the 20th century, Deliverable D9 of STARDEX project: STAtistical and Regional dynamical Downscaling of EXtremes for European regions (www.stardex-eu.org)

Barnett TP, Adam JC, Lettenmaier DP (2005) Potential impacts of a warming climate on water availability in snow-dominated regions. Nature 438(7066):303–309

Bavay M, Lehning M, Jonas T, Lowe H (2009) Simulations of future snow cover and discharge in Alpine headwater catchments. Hydrol Process 23(95–108):10. doi:10.1002/hyp.7195

Bednorz E (2002) Snow cover in western Poland and macro-scale circulation conditions. Int J Climatol 22:533–541. doi:10.1002/joc.752

Beniston M (ed) (1994) Mountain environments in changing climates. Routledge, London/New York, p 492

Beniston M (1997) Variations of snow depth and duration in the Swiss Alps over the last 50 years: links to changes in large-scale climatic forcings. Clim Change 36(3–4):281–300

Beniston M (2000) Les multiples contraintes environnementales menant aux risques naturels en régions de montangne, Actes du 5[e] Colloque transfrontalier CLUSE, Université de Genève, Geneva

Beniston M (2005) Mountain climates and climatic change: an overview of processes focusing on the European Alps. Pure Appl Geophys 162(2005):1587–1606

Beniston M (2006) Mountain weather and climate; a general overview and a focus on climatic change in the Alps. Hydrobiologia 562:3–16

Beniston M, Rebetez M (1996) Regional behavior of minimum temperatures in Switzerland for the period 1979–1993. Theor Appl Climatol 53:231–243

Beniston M, Rebetez M (1997) Extreme climatic events and variability. In: conference on Global Climate Observing Systems (GCOS), June 3–6, Asheville, NC, USA

Beniston M, Diaz HF, Bradley RS (1997) Climatic change at high elevation sites: an overview. Clim Change 36:233–252

Beniston M, Keller F, Koffi B et al (2003) Estimates of snow accumulation and volume in the Swiss Alps under changing climatic conditions. Theor Appl Climatol 76:125–140

Birsan MV, Dumitrescu A (2014) Snow variability in Romania in connection to large-scale atmospheric circulation. Int J Climatol 34:134–144. doi:10.1002/joc.3671

Birsan MV, Zaharia L, Chendes V, Branescu E (2012) Recent trends in streamflow in Romania (1976–2005). Rom Rep Phys 64(1):275–280
Birsan MV, Marin L, Dumitrescu A (2013) Seasonal changes in wind speed in Romania. Rom Rep Phys 65(4):1479–1484
Birsan MV, Zaharia L, Chendes V, Branescu E (2014) Seasonal trends in Romanian streamflow. Hydrol Process 28(15):4496–4505. doi:10.1002/hyp.9961
Bodin X, Thibert E, Fabre D, Ribolini A, Schoeneich P, Francou B, Reynaud L, Fort M (2009) Two decades of responses (1986–2006) to climate by the Laurichard Rock Glacier, French Alps. Permafr Periglac Process 20(4):331–344
Bogdan O, Marinică I, Rusan N, Rusu S (2007) Riscul iernilor calde în România (cu aplicații la iarna 2006–2007). Riscuri și Catastrofe, VI 4:97–110
Böhm R, Auer I, Brunetti M, Maugeri M, Nanni T, Schoner W (2001) Regional temperature variability in the European Alps: 1760–1998 from homogenized instrumental time series. Int J Climatol 21:1779–1801
Bojariu R, Dinu M (2007) Snow variability and change in Romania. In: Proceedings of Alpine Snow workshop, Forschungsbericht, 53, pp. 34–38
Bojariu R, Paliu D (2001) North Atlantic oscillation projection on Romanian climate fluctuations in the cold season. In: Brunet M, Lopez D (eds) Detecting and modelling regional climate change and associated impacts. Springer-Verlag, Berlin, pp 345–356
Brown RD, Petkova N (2007) Snow cover variability in Bulgarian mountainous regions, 1931–2000. Int J Climatol 27:1215–1229. doi:10.1002/joc.1468
Busuioc A, Caian M, Cheval S, Bojariu R, Boroneant C, Baciu M, Dumitrescu A (2010) Variabilitatea și schimbarea climei în România. ProUniversitaria, București, p 226
Cayan DR, Kammerdiener S, Dettinger MD, Caprio JM, Peterson DH (2001) Changes in the onset of spring in the western United States. Bull Am Meteorol Soc 82(3):399–415
Ciccarelli N, von Hardenberg J, Provenzale A, Ronchi C, Vargiu A, Pelosini R (2008) Climate variability in north-western Italy during the second half of the 20th century. Global Planet Change 63:185–195
Davis ED, Lowit MB, Knappenberger PC (1999) A climatology of snowfall temperature relationships in Canada. J Geophys Res 104(D10):11985–11994. doi:10.1029/1999JD900104
Diaz HF, Bradley RS (1997) Temperature variations during the last century at high elevation sites. Clim Change 36:253–279
Diaz HF, Grosjean M, Graumlich L (2003) Climate variability and change in high elevation regions: past, present and future. Clim Change 59:1–4
Dima V (2012) Formațiuni anticiclonice Euro-Atlantice și fenomene meteorologice de risc din estul și sud-estul României, Teză de doctorat (mss). Institutul de Geografie, București
Dye DG (2002) Variability and trends in the annual snow-cover cycle in Northern Hemisphere land areas. 1972–2000. Hydrol Process 16:3065–3077
Dyurgerov M (2003) Mountain and subpolar glaciers show an increase in sensitivity to climate warming and intensification of the water cycle. J Hydrol 282:164–176
Easterling DR (2002) Recent changes in frost days and the frost-free season in the United States. Bull Am Meteorol Soc 83:1327–1332
EEA (2008) Impacts of Europe's changing climate – 2008 indicator-based assessment. Report no. 4. Joint EEA-JRC-WHO report. p 19
EEA (2010) Europe's ecological backbone: recognising the true value of our mountains. Report no. 6. p 248
Etchevers P, Martin E (2002) L'eau en montagne. Impact d'un changement climatique sur le manteau neigeux et l'hydrologie des bassins versants de montagne. Centre d'Etudes de la Neige, Météo France, Dordrecht (Springer Netherlands) p 10
Fagre DB, Peterson DL, Hessl AE (2003) Taking the pulse of mountains: ecosystem responses to climatic variability. Clim. Chang 59: 263–282; Fărcaş. I. (1981). Contributions au problème de la variation annuelle de la radiation solaire directe sur le surfaces avec différentes orientations et inclinations aux latitudes moyennes. RRGGG-Géogr. 25:2

Frei C, Schär C (2001) Detection probability of trends in rare events: theory and application to heavy precipitations in the alpine region. J Climate 14:1568–1584

Global outlook for ice and snow (2007) United Nations Environment Programme, Division of Early Warning and Assessment (DEWA), Nairobi, p 235

Grabherr G (2003) Overview: alpine vegetation dynamics and climate change – a synthesis of long-term studies and observations, in: alpine biodiversity in Europe. Springer, Berlin

Groisman P, Karl TR, Knight RW, Stenchikov G (1994) Changes of snow cover, temperature and radiative heat balance over the northern hemisphere. J Climate 7:1633–1656

Groisman PY, Knight RW, Easterling DR, Karl TR, Hegerl GC, Razuvaev VN (2005) Trends in intense precipitation in the climate record. J Climate 18:1326–1350

Grunewald K, Scheithauer J, Monget J-M, Brown D (2009) Characterisation of contemporary local climate change in the mountains of southwestern Bulgaria. Clim Change 95(3–4):535–549. doi:10.1007/s10584-008-9508-8

Hennegriff W, Kolokotronis V, Weber H et al (2006) Climate change and floods – findings and adaptation strategies for flood protection. KA Abwasser Abfall 53:8

Huntington TG (2006) Evidence for intensification of the global water cycle: review and synthesis. J Hydrol 319:83–95. doi:10.1016/j.jhydrol.2005.07.003

IPCC (2007) Climate change 2007: Contribution of Working Group I to the fourth assessment report of the Intergovernmental Panel on Climate Change (Solomon S, Qin D, Manning M, Chen Z, Marquis M, Averyt KB, Tignorand M, Miller HL, eds) Cambridge University Press, Cambridge/New York

IPCC (2013) Climate change 2013: the physical science basis. Contribution of Working Group I to the fifth assessment report of the Intergovernmental Panel on Climate Change (Stocker TF, Qin D, Plattner G-K, Tignor M, Allen SK, Boschung J, Nauels A, Xia Y, Bex V, Midgley PM, eds) Cambridge University Press, Cambridge/New York

Jomelli V, Pech P, Chochillon C, Brunstein D (2004) Geomorphic variations of debris flows and recent climatic change in the French Alps. Clim Change 64:77–102

Jungo P, Beniston M (2001) Changes in the anomalies of extreme temperature anomalies in the 20th century at Swiss climatological stations located at different latitudes and altitudes. Theor Appl Climatol 69:1–12

Karl TR, Jones PD, Knight RW, Kukla G, Plummer N, Razuvayev V, Gallo KP, Lindseay J, Charlson RJ, Peterson TC (1993) A new perspective on recent global warming: asymmetric trends of daily maximum and minimum temperature. Bull Am Meteorol Soc 74:1007–1023

Karl TR, Knight RW, Easterling DR, Quayle RG (1996) Indices of climate change for the United States. Bull Am Meteorol Soc 77:279–291

Keller F (2000) Les possibilités de changements dans la couverture végétale alpine peuvent-elle être condérées des risques, Actes du 5[e] Colloque transfrontalier CLUSE, Université de Genève

Keller F, Kienast F, Beniston M (2000) Evidence of response of vegetation to environmental change on high-elevation sites in the Swiss Alps. Reg Environ Change 1(2):70–77

Klein Tank A (2004) Changing temperature and precipitation extremes in Europe's climate of the 20th century. PhD-thesis, De Bilt, p 124. Available on http://eca.knmi.nl/documents/Thesis_KleinTank_2004.pdf

Koleva M, Iotova AM (1992) Snow cover variability and change in mountainous regions of Bulgaria, 1931–2000. Meteorol Z 13(1):19–23

Küttel M, Luterbacher J, Wanner H (2011) Multidecadal changes in winter circulation-climate relationship in Europe: frequency variations, within-type modifications, and long-term trends. Clim Dyn 36(5–6):957–972. doi:10.1007/s00382-009-0737-y

Laternser M, Schneebeli M (2003) Long-term snow climate trends of the Swiss Alps (1931-99). Int J Climatol 23:733–750

Li J, Wang J (2003) A new North Atlantic oscillation index and its variability. Adv Atmos Sci 20 (5):661–676. doi:10.1007/BF02915394

Martin E, Etchevers P (2005) Impact of climatic changes on snow cover and snow hydrology in the French Alps, global change and mountain regions. Adv Global Chang Res 23:235–242, Part II

Martin E, Giraud G, Lejeune Y, Boudart G (2001) Impact of climate change on avalanche hazard. Ann Glaciol 32:163–167

McVicar TR, Roderick ML (2010) Atmospheric science: winds of change. Nat Geosci 3:747–748. doi:10.1038/ngeo1002

McVicar TR et al (2008) Global review and synthesis of trends in observed terrestrial near-surface wind speeds: implications for evaporation. J Hydrol 416–417:182–205. doi:10.1016/j.jhydrol.2011.10.024

Messerli B, Ives JD (eds) (1997) Mountains of the world: a global priority. Parthenon, New York, p 495

Micu D (2009) Snow pack in the Romanian Carpathians under changing climatic conditions. Meteorol Atmos Phys 105(1–2):1–16

Nogués-Bravo D, Araújo MB, Erread MP, Martinez-Ricad JP (2007) Exposure of global mountain systems to climate warming during the 21st century. Glob Environ Chang 17:240–428

Nojarov O (2012) Variations in precipitation amounts, atmosphere circulation, and relative humidity in high mountainous parts of Bulgaria for the period 1947–2008. Theor Appl Climatol 107:175–187

ONERC (Observatoire National sur les Effets du Réchauffement Climatique) (2007) Un climat à la dérive: comment s'adapter ? Rapport au Premier ministre et au Parlement. La documentation Française, Paris, p 107

Petkova N, Alexandrov V (2012) Snow cover in Bulgarian mountainous regions: observed variations and future activities, BALWOIS 2012

Petkova N, Koleva E, Alexandrov V (2004) Snow cover variability and change in mountainous regions of Bulgaria, 1931–2000. Meteorol Z 13(1):19–23

Petkova N, Alexandrov V, Koleva E (2010) Snow cover variability in Bulgaria, 1931–2005. In: Proceedings of the international conference: geography and regional development, 14–16 October, 2010

Schneebeli M, Laternser M, Ammann W (1997) Destructive snow avalanches and climate change in the Swiss Alps. Eclogae Geol Helv 90:457–461

Schöner W, Auer I, Böhm R (2009) Long term trend of snow depth at Sonnblick (Austrian Alps) and its relation to climate change. Hydrol Process 23:1052–1063

Schröter D, Cramer W, Leemans R, Araújo Arnell NW, Bondeau A, Bugmann H, Carter TH (2005) Ecosystem service supply and vulnerability to global change in Europe. Science 310:1333–1337. doi:10.1126/science.1115233

Seiler, W. (2006), Les changements climatiques dans l'espace alpin: tendances, retombées et defies, in: Deuxième manifestation thématique "Changement du climat dans l'espace alpin – Effets et défis" à l'occasion de la 31e réunion du Comité permanent à Galtür, Wien, pp. 7–19

Spehn EM, Messerli B, Körner C (2002) A global assessment of mountain biodiversity: synthesis. In: Körner C, Spehn EM (eds) Mountain biodiversity: a global assessment. The Parthenon Publishing Group, London

Stehr N, von Storch H (2010) Climate and society: climate as resource, climate as risk. World Scientific Publishing, Hackensack, 200 p

Valt M, Cagnati A, Crepaz A, Marigo G (2005) Snow precipitation in the last years in the Italian Alps. Paper presented at 28th international conference on Alpine Meteorology (ICAM) and the annual scientific meeting of the Mesoscale Alpine Programme (MAP) 2005, Meteorol. and Hydrol. Serv. of Croatia, Zadar, Croatia

Velev S (1996) Is Bulgaria become warmer and drier. GeoJournal 40(4):363–370

Wanner H, Brönnimann S, Casty C, Gyalistras D, Luterbacher J, Schmutz C, Stephenson DB, Xoplaki E (2001) North Atlantic Oscillation – concepts and studies. Surv Geophys 22:321–382. doi:10.1023/A:1014217317898

Wilson RJ, Gutierrez D, Gutierrez J, Martinez D, Agudo R, Monserrat VJ (2005) Changes to the elevational limits and extent of species ranges associated with climate change. Ecol Lett 8:1138–1146

WMO (2011) WMO statement on the status of the global climate in 2010. WMO-No. 107. World Meteorological Organization, Geneva, p 20

Woodwell GM (2004) Mountains: top down. Ambio Suppl 13:35–38

Żmudzka E (2010) Changes in thermal conditions in the high mountain areas and contemporary warming in the central Europe. Misc Geogr 14:59–70

Żmudzka E (2011) Contemporary climate changes in the high mountain part of the Tatras. Misc Geogr 15:93–102

# Chapter 9
# Changing Climate Extremes in the Last Five Decades (1961–2010)

**Abstract** Several indices of climate extremes – based on minimum and maximum temperature, daily precipitation and daily snow cover – were used in order to check for changes over 1961–2010. The most important changes were found in maximum and (to a lesser extent) in minimum seasonal temperature. The warming signal is well retrieved in the trends in thermal-related extreme indices. Autumn is the only stable season with respect to changes in temperature extremes. Precipitation extremes exhibit no consistent change, leading to the conclusion that the decreasing trends in the snow-related indices, especially the maximum length of snowfall spells are rather related to recent warming.

Mountain regions are particularly sensitive to warming and drying, associated with more frequent and intense climate extremes (e.g. heat spells, cold spells, heavy rainfalls, heavy snowfalls) most visible in the last decades of the twentieth century (Keller 2000; Messerli and Ives 1997; Beniston et al. 1997; Beniston and Rebetez 1996; Diaz and Bradely 1997; Pepin and Seidel 2005; IPCC 2007). The public and scientific community changed their view of likely extremes and their awareness has risen significantly in the recent years. There was a great emphasis on the effects and losses of some headline climate extreme events in Europe [e.g. the heat wave of summer 2003, the Anatol, Lothar and Martin storms, the floods of 2002]. Sets of climate extreme indices have been developed through several international workshops (Folland et al. 1999) and EU-financed projects (STARDEX, EMULATE, MICE, PRUDENCE) to enable global analyses, comparisons between regions with different climate regimes and to establish potential relationships between isolated cases of extremes and long-term climate trends (Frei and Schär 2001). International coordinated work of CCL/CLIVAR/JCOMM Expert Team of Climate Change Detection (ETCCDI) have been conducted and provided 27 core climate extreme indices and special software for their computation based on observational data. These indices proposed by the European Climate Assessment and Dataset (ECA&D), an early European initiative in this view, and ETCCDI largely correspond.

In this work, the detection of changes relative to the variability of climate extremes in the Carpathians has been made by using the WMO-CCL/CLIVAR guidelines. This chapter summarizes the change signals in temperature, precipitation and snow extremes in these mountains. Statistical trends evaluated

 
D.M. Micu et al., *Climate of the Romanian Carpathians*,
Springer Atmospheric Sciences, DOI 10.1007/978-3-319-02886-6_9

for 24 climate extreme indices selected from the core set indices proposed by ETCCDI, have been analyzed and compared regionally (if any patterns observed) and by elevation levels over a WMO representative period covering the last five decades, relative to the 1961–1990 WMO baseline. Table 9.1 presents the definitions and parameters of the climate extreme indices selected in this study.

**Table 9.1** List of the indices of climate extremes analyzed in this study

| Acronym | Description | Definition | Unit |
|---|---|---|---|
| Temperature extremes | | | |
| Tn10p | Cold nights | Percent of time when $T_{min}$ <10th percentile | % |
| Tn90p | Warm nights | Percent of time when $T_{min}$ >90th percentile | % |
| Tx10p | Cold nights | Percent of time when $T_{max}$ <10th percentile | % |
| Tx90p | Cold nights | Percent of time when $T_{min}$ >90th percentile | % |
| GSL | Growing season length | Count of days between first span of at least 6 days $T_{avg}$ >5 °C and first span in second half of the year of 6 days $T_{avg}$ <5 °C | Days |
| WSDI | Warm spell duration index | The number of days each year which are part of a warm Spell. A warm spell is defined as a sequence of 6 or more days in which the daily maximum temperature exceeds the 90th percentile of daily maximum temperature for a 5-day running window surrounding this day during the baseline period | Days |
| CSDI | Cold spell duration index | The number of days each year which are part of a cold spell. A cold spell is defined as a sequence of 6 or more days in which the daily minimum temperature is below the 10th percentile of daily minimum temperature for a 5-day running window surrounding this day during the baseline period | Days |
| FD | Frost days | Annual sum of days with daily minimum temperature below 0 °C | Days |
| ID | Icing days | Annual sum of days with daily maximum temperature below 0 °C | Days |
| SU | Summer days | Annual sum of days with daily maximum temperature above 25 °C | Days |
| TR | Tropical nights | Annual sum of days with daily minimum temperature above 20 °C | Days |
| Precipitation extremes | | | |
| CDD | Dry spells | Maximum number of consecutive dry days (<1.0 mm) | Days |
| CWD | Wet spells | Maximum number of consecutive wet days (>1.0 mm) | Days |

(continued)

**Table 9.1** (continued)

| Acronym | Description | Definition | Unit |
|---|---|---|---|
| R10 | Days with over 10 mm of precipitation | Annual sum of days where daily precipitation is more than 10 mm | Days |
| R20 | Days with over 20 mm of precipitation | Annual sum of days where daily precipitation is more than 20 mm | Days |
| R95PTOT | Heavy precipitation days | Annual sum of precipitation in days where daily precipitation exceeds the 95th percentile of daily precipitation in the baseline period | mm |
| R99PTOT | Very heavy precipitation days | Annual sum of precipitation in days where daily precipitation exceeds the 99th percentile of daily precipitation in the baseline period | mm |
| PRCPTOT | Precipitation amount | Annual sum of precipitation in days where daily precipitation is at least 1 mm/day | mm |
| SDII | Simply precipitation intensity index | Quotient of precipitation amount in wet days (>1.0 mm) and number of wet days | mm/day |
| Snow extremes | | | |
| HSS | Snowfall spells | Maximum number of consecutive snowfall days | Days |
| HS1day<br>H2day<br>H3day | Greatest 1-, 2-, and 3-day snow depths, respectively | Maximum seasonal snow depths cumulated over 1–3 consecutive days | cm |
| SD50 cm | Consecutive days of abundant snow accumulation | Maximum duration of thick snowpack intervals (≥50 cm) | Days |

## 9.1 Change in Air Temperature Extremes

Prior works examining the variability of temperature extremes over the twentieth century in different mountain regions confirmed the increase in the number of days with extremely high temperatures and the tendency towards fewer days with extremely low temperatures (Baeriswyl and Rebetez 1996; Beniston and Rebetez 1997; Weber et al. 1994, 1997). Analyzing the variability of climate extremes in the Romanian Carpathians (above 1,000 m) over 1961–2003, Cheval et al. (2004) provided reliable evidence of an ongoing warming and emphasized the intensification of temperature and precipitation extremes (e.g. more warm extremes and less cold ones, higher number of consecutive dry days).

The spatial patterns of seasonal trends in minimum and maximum temperature during 1961–2010 are presented in Figs. 9.1 and 9.2, respectively. Similar to the signal shown by the mean temperature trends, the seasonal temperature extremes present exclusively warming trends in all seasons except autumn, with fewer statistically significant trends in minimum temperature.

The inter-annual variability of selected percentile-based indices suggests an intensification of warm extremes mostly in medium elevation areas (generally

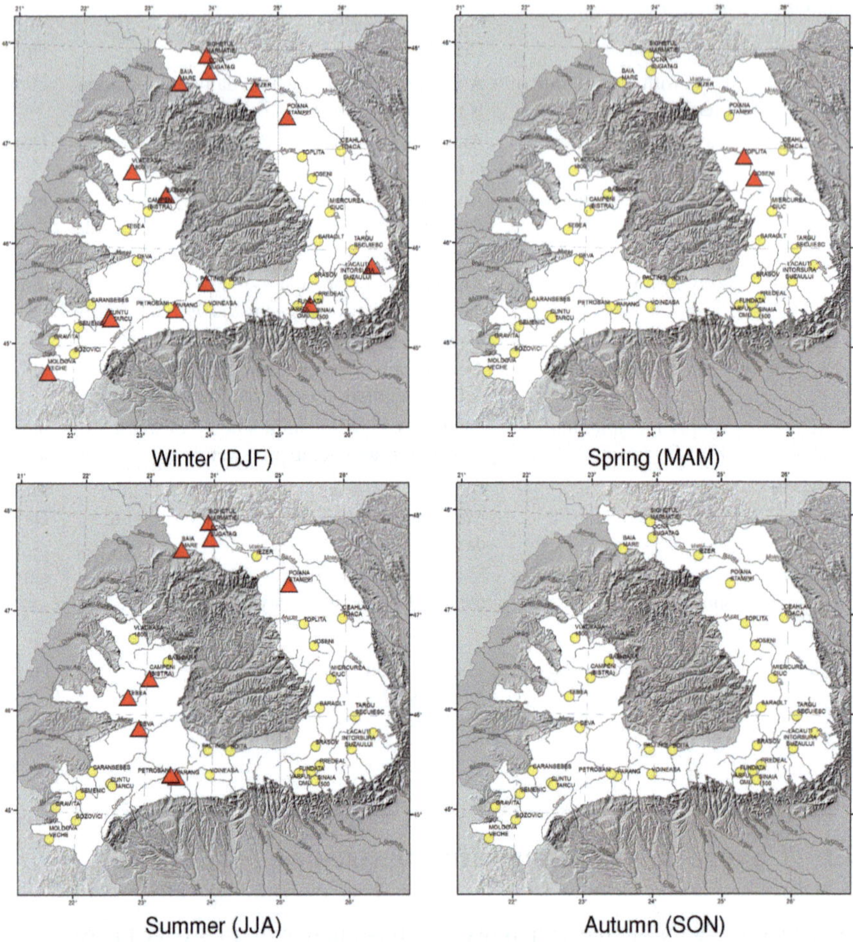


**Fig. 9.1** Seasonal trends in minimum air temperature. Significant increasing trends are symbolized with *upward red triangles*

below 1,800 m). There are no important changes in the variability of Tn10p and Tx10p. The variability of temperature extremes defined by the 90th percentile outlines a progressive and widespread warming in most areas of the Carpathians. The frequency of *warm nights* (Tn90p) and *warm days* (Tx90p) is a valuable temperature change indicator in the mountain regions. Under the influence of altitude which moderates the heating process, the frequency of *warm nights* (Tn90p) in the areas above 800 m is 20–40 %. High night-time $T_{min}$ values, ranging from 4 to 11 °C, present significant and widespread increases across the Carpathians. The signal of Tn10p intensification is stronger since mid or late 1980s at several medium elevation sites of the Southern and Western Carpathians and exceptionally, at about 2,200 m (the alpine weather stations). The values of 90th percentile depicting the incidence of *warm days* are 8–11 °C lower than the

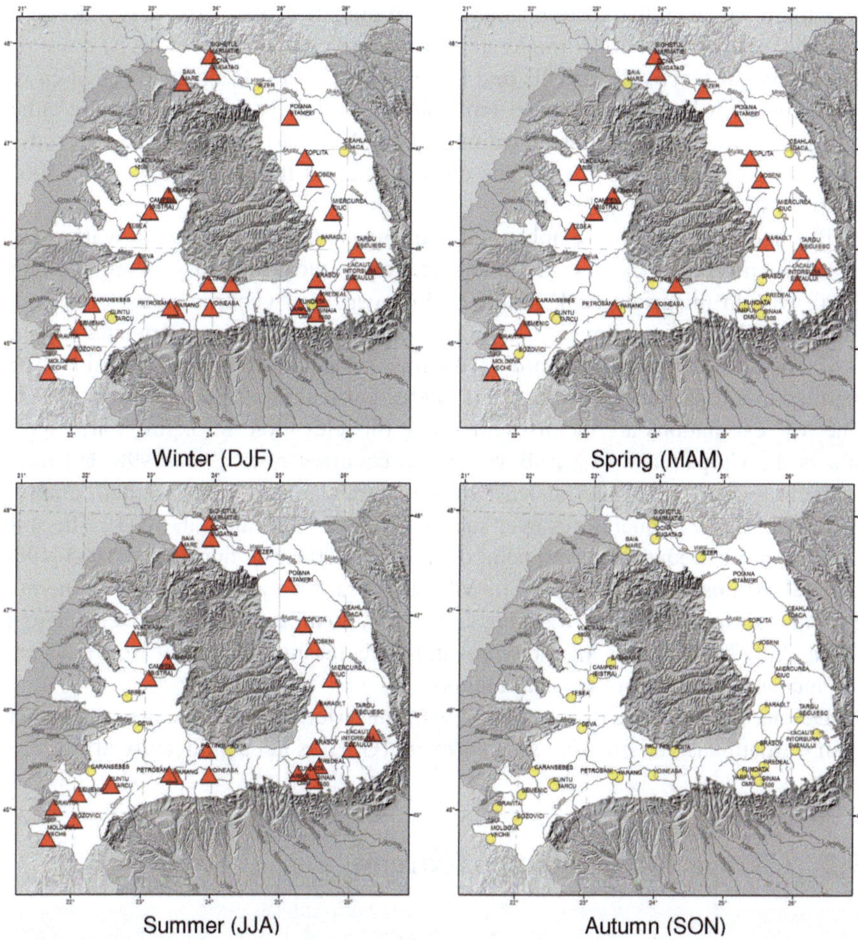


**Fig. 9.2** Seasonal trends in maximum air temperature. Significant increasing trends are symbolised with *upward red triangles*

$T_{max}$ absolute records until 2003. The MK test revealed consistent upward trends of Tx90p, highly significant at medium elevations of the Southern and Western Carpathians, and exceptionally, at the highest elevation station of Eastern Carpathians (for about half of the stations). The shifting years depicting the intensification of warming cover the 1985–1999 interval (but mostly the 1985–1987) and they are fairly comparable to those observed at country scale (Busuioc et al. 2010).

*Cold spells* are assumed to be a suggestive indicator of negative temperature extremes, often associated with massive cooling spells, frequently growing into absolute negative thermal records. Busuioc et al. (2010) asserted that cold waves in Romania last less than heat waves. Regionally, it appears that the average length of cold spells is the shortest in the Western Carpathians (about 9.9 days), with maximum values in the Eastern and Southern branches (about 10.5 days). Regardless of the altitude, short or moderately persistent cold spells (below 10 days and 11–30 days, respectively) are the best represented (50–75 % and 20–45 % of cases, respectively), while the very persistent cold spells (longer than 30 days) are rather rare and characteristic of high elevations (above 1,800 m). Longest cold spells after 1961 were generally recorded in the coldest years of the period (e.g. 1962–1963, 1980, 1985, 1987, 1991, 1997).

Busuioc et al. (2010) emphasized the outstanding evolution of warm spell duration and frequency towards intensification in most regions of Romania after 2000 with an important peak in 2007. However, their occurrence is not entirely linked to 'hot' climate regions of this country (e.g. the southern and eastern lowlands). The transition towards a warmer mountain climate since mid 1980s resulted in a higher incidence of warm spells, particularly in areas below 1,600 m. The MK estimations are inconsistent above this elevation. Strongest warm spells across the Carpathians ($T_{max}$ above 25 °C) occurred after 1989–1990, but more frequent since 2000, when at least five-six mountain stations recorded simultaneously extreme temperatures over successive days (Cheval et al. 2004). The shift towards more frequent warm spells over 1961–2003 was during the 1985–1988 interval at most sites, according to MK statistics. Overall, trends in the thermal extremes show a consistent warming signal (i.e., no mixed trends), as it can be seen in Fig. 9.3. The signal is most obvious in the decreasing number of frost days and the increasing number of tropical nights.

Despite the evidence of an ongoing warming over extended areas in the Romanian Carpathians, the *growing season length* (GSL) is not significantly affected.

## 9.2 Change in Precipitation Extremes

Mountain regions are characterized by complex spatial and temporal precipitation patterns. The trend signals of precipitation extremes are not homogenous or consistent across the Carpathians, neither regionally nor by elevation. However, the dry extremes prevail at most sites located above 1,000 m, fitting the general annual and seasonal drying trends observed in these mountains.

The seasonal maximum daily precipitation presents mixed trends at less than 20 % of the stations in all seasons except autumn, when the significant trends are exclusively increasing (Fig. 9.4).

The length of *wet spells* (CWD) shows a tendency to diminish only at three sites (Baraolt, Ceahlău Toaca and Lăcăuți), while the *length of dry spells* (CDD) remains unchanged except at Toplița, where it presents a significant downward trend. Overall, it can be stated that both indices are stable.

Trend maps in other six of the precipitation-related indices of extremes are shown in Fig. 9.5. The frequency of days with precipitation over 10 and 20 mm (R10 and R20) present a very similar pattern, with increasing trends in western and northwestern parts of the study area.

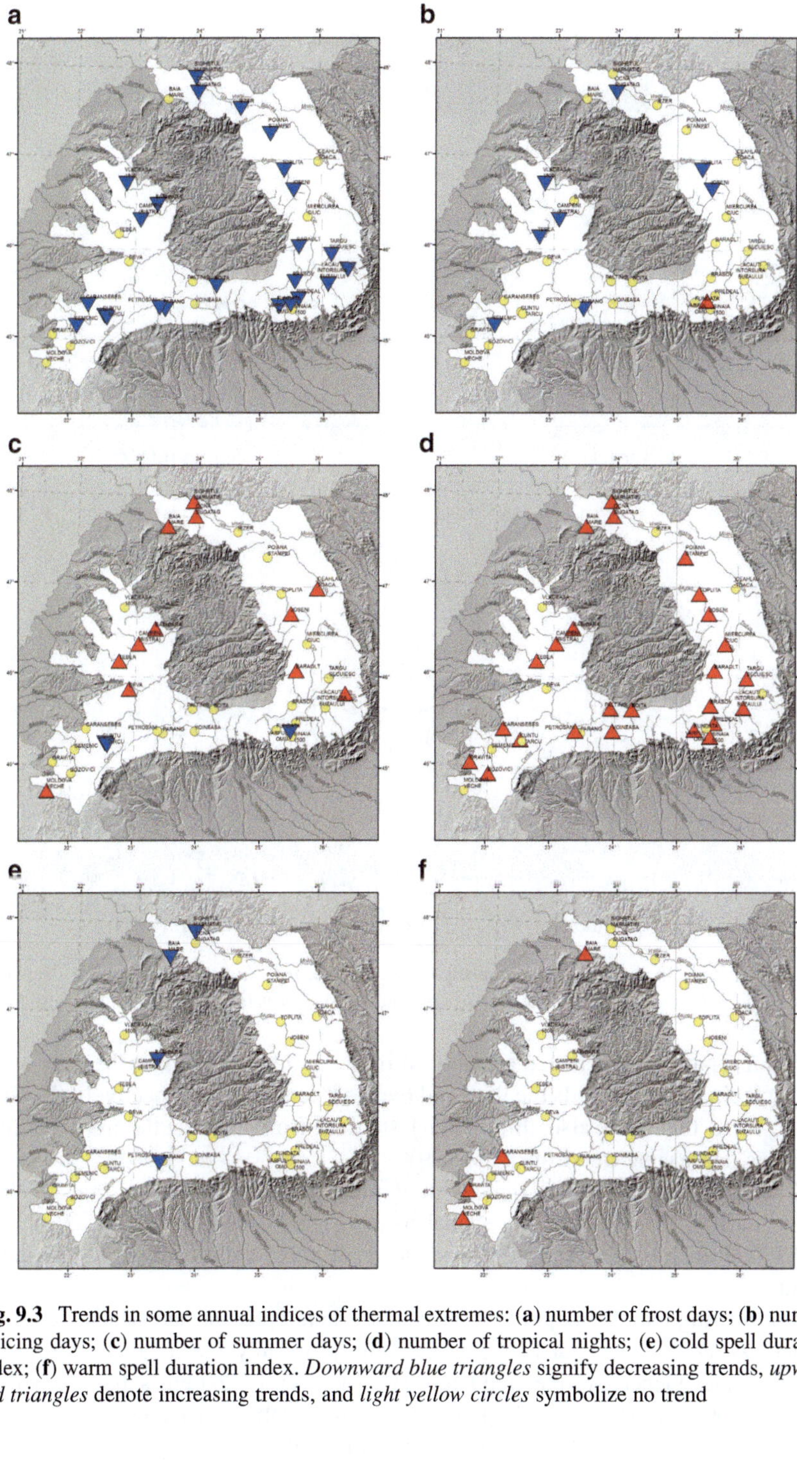


**Fig. 9.3** Trends in some annual indices of thermal extremes: (**a**) number of frost days; (**b**) number of icing days; (**c**) number of summer days; (**d**) number of tropical nights; (**e**) cold spell duration index; (**f**) warm spell duration index. *Downward blue triangles* signify decreasing trends, *upward red triangles* denote increasing trends, and *light yellow circles* symbolize no trend

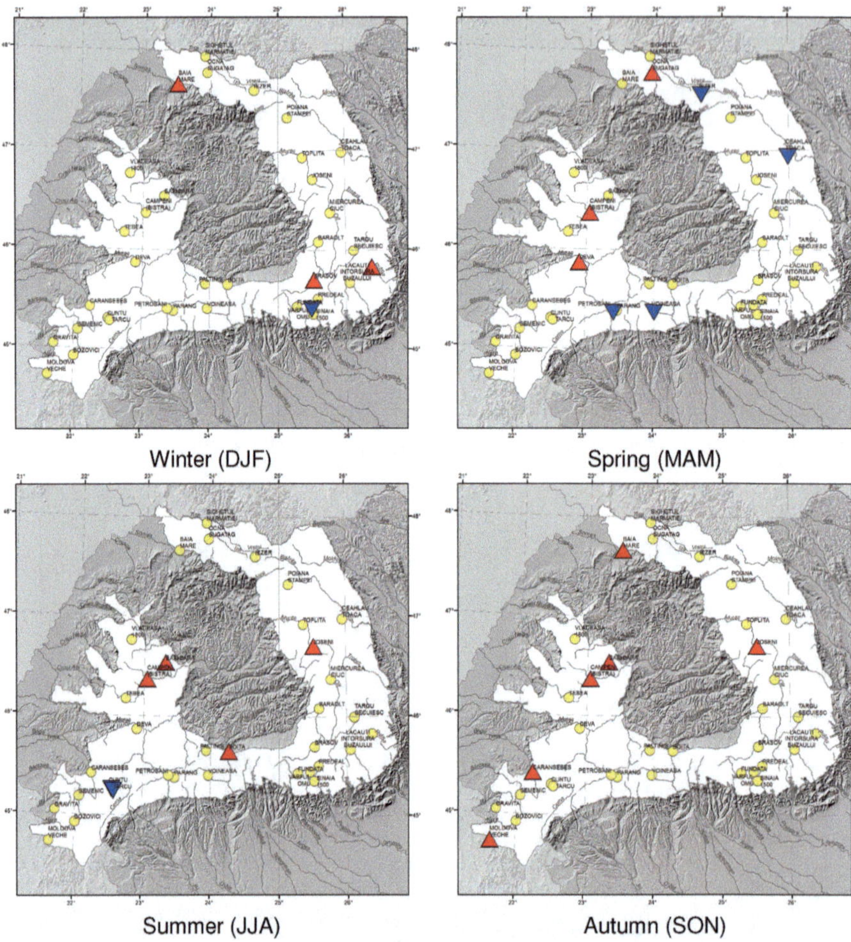


**Fig. 9.4** Trends in maximum daily precipitation. *Downward blue triangles* signify decreasing trends, *upward red triangles* denote increasing trends, and *light yellow circles* symbolize no trend

Similar patterns of change are noticed in the number of days of the year where daily precipitation exceeds the 95th and the 99th percentiles, respectively (R95 and R99). Here there are also three (four) stations from the Southern Carpathians showing decreasing trends in R95 (R99).

The SDII and PRCPTOT indices show predominantly increasing trends. While trends in SDII do not show any spatial patterns across the Carpathians, changes in PRCPTOT are a little bit more grouped; however, most station present no significant trend.

Considering that the majority of the stations do not present statistically significant trends, we could state that long-term variability of daily precipitation extremes is rather stable. Generally, the trend results of indices of extremes based on daily precipitation over the 1961–2010 period do not support the idea that in the Carpathians under a warmer climate, higher daily precipitation intensity events are more likely.

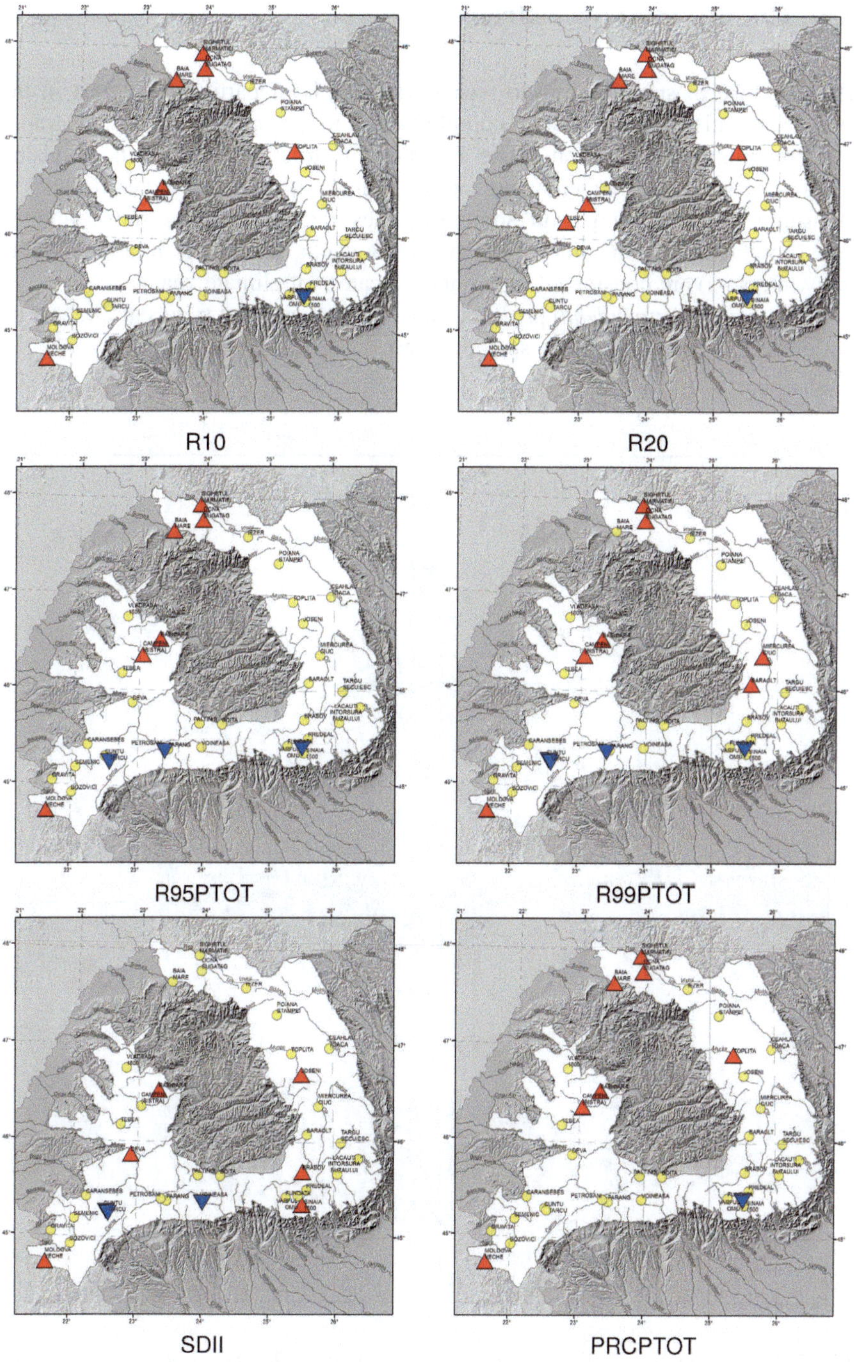


**Fig. 9.5** Trends in precipitation-related indices of extremes. *Downward blue triangles* signify decreasing trends, *upward red triangles* denote increasing trends, and *light yellow circles* symbolize no trend

## 9.3 Change in Snow Extremes

The response of snow regime to recent warming is of particular interest in this work, since this variable is rather complex and it may respond both to temperature and precipitation oscillations. The Romanian Carpathians experienced visible declines of snowfall frequency and snow cover duration over the last decades of the study period. Changes in some extremes related to large snow amounts (e.g. snowfall spells, maximum seasonal snow depth, maximum duration of thick snow intervals) are considered important indicators of climate change.

The *maximum snow depth* presents statistically significant trends at 25 % of the stations (Fig. 9.6), showing no spatial pattern.

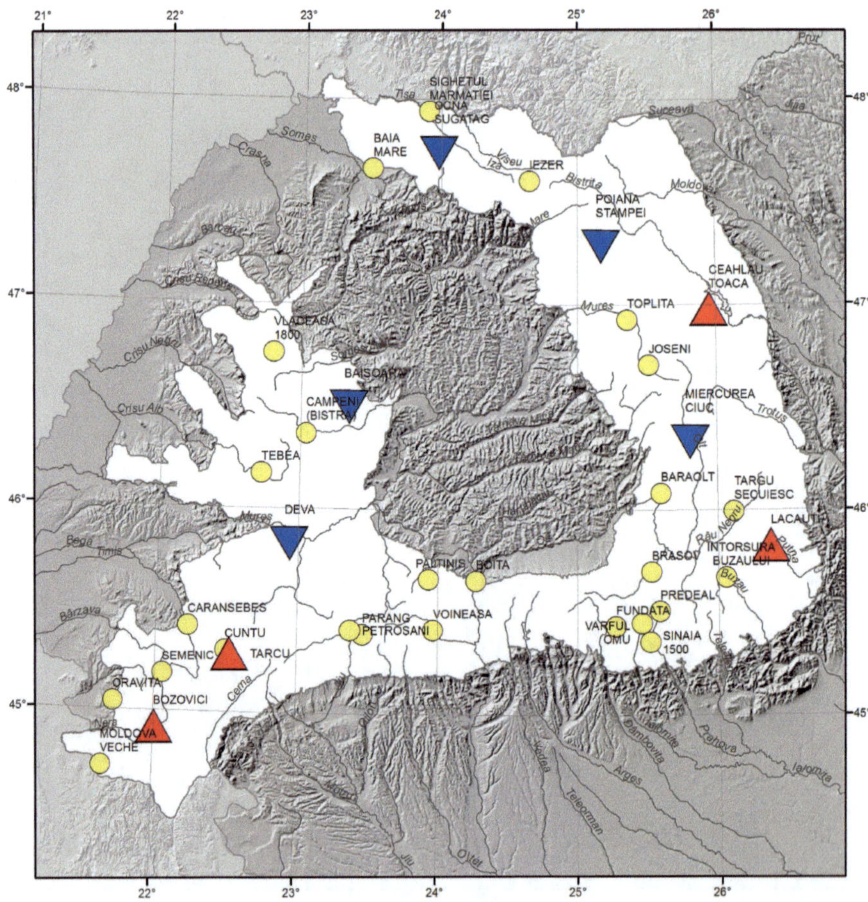


**Fig. 9.6** Trends in precipitation-related indices of extremes. *Downward blue triangles* signify decreasing trends, *upward red triangles* denote increasing trends, and *light yellow circles* symbolize no trend

*The maximum length of snowfall spells* (HSS) in the Carpathians exceeded rarely 16 days, particularly during 1970s. There is an obvious downward trend in the evolution of this index since 1980s all over Carpathians, rather strong in mid-elevation areas. At these elevations HSS reached up to 12–14 days after 1990. At about 2,500 m (Vf. Omu), change signal in HSS became consistent since mid 1980s. No regional change patterns were observed in the HSS variability. At most stations, HSS is on a significant decrease at a rate of maximum 2 days/decade corresponding to the general trends described by seasonal snowfalls and precipitation.

The winter-to-winter variability of *maximum snow depth cumulated over short-time spans* (HS1-3day) is rather high and trends towards snow abundance is limited to few areas of the Southern (Păltiniş) and Eastern Carpathians (Lăcăuţi). Most changes associated with this index correspond to an average decrease of 3–6 cm/decade for HS1day and 19 cm/decade for HS3day (only at Băişoara). However, positive trends revealed by MK test suggest stronger change rates associated to the variability of these extremes, yet sparsely distributed in these mountains. The change signals in the maximum snow depth variability became obvious since early 1980s, particularly in the areas subject to consistent upward trends. Generally, there is good agreement between the signals in maximum snow accumulation indices and those depicted by the trends of heavy snowfalls.

The *number of days with thick snow* (SD50cm) is of more practical use than other snow extreme indices, depicting proper conditions for skiing and snowboarding in mid and high-elevation areas. The days with snow depth over 50 cm prevail over the year in all the areas above 1,800 m of the Romanian Carpathians. A higher variability has been observed at low-laying weather stations, where this threshold is more rarely maintained over several successive days. Lowest spells of SD50cm were of only 2–7 days duration. There are no consistent long-term changes observed over extended areas of the Carpathians for this index. Generally, the areas subject to significant more snow abundant-related extremes tend to experience longer intervals of thick snow (with 2.7 days/decade at Lăcăuţi, Eastern Carpathians), while the opposite is true for the areas where a lower incidence of short duration snow accumulations has been estimated (1.72 days/decade at Băişoara, Western Carpathians). These changes were observed since early 1980s.

The findings of this chapter support the idea that a warmer climate in the Romanian Carpathians has resulted both in higher and lower extreme snow accumulations, but these trends are only sparsely observed across the Carpathians. The general background of the changes in snow extremes is given by a widespread and consistent tendency towards more frequent snow showers, but shorter snowfall spells (e.g., Birsan and Dumitrescu 2014).

## References

Baeriswyl PA, Rebetez M (1996) Répartition spatiale des modifications climatiques dans le domain alpin. Swiss National Science Foundation. NFP 4031-038270. p 236

Beniston M, Rebetez M (1996) Regional behavior of minimum temperatures in Switzerland for the period 1979–1993. Theor Appl Climatol 53:231–243

Beniston M, Rebetez M (1997) Extreme climatic events and variability, conference on Global Climate Observing Systems (GCOS), June 3–6, Asheville

Beniston M, Diaz HF, Bradley RS (1997) Climatic change at high elevation sites: an overview. Clim Change 36:233–252

Birsan MV, Dumitrescu A (2014) Snow variability in Romania in connection to large-scale atmospheric circulation. Int J Climatol 34:134–144. doi:10.1002/joc.3671

Busuioc A, Caian M, Cheval S, Bojariu R, Boroneanţ C, Baciu M, Dumitrescu Al (2010) Variabilitatea şi schimbarea climei în România (Climate variability and change in Romania). Edit. ProUniversitaria, Bucureşti. p 226 (in Romanian)

Cheval S, Baciu M, Breza T (2004) The variability of extreme events in the Romanian Carpathians. AUVT-Geogr XIV:59–78

Diaz HF, Bradley RS (1997) Temperature variations during the last century at high elevation sites. Clim Change 2:21–47

Folland CK, Horton EB, Scholefield P (eds) (1999) Report of WMO Working Group on Climate Change. Detection Task Group on Climate Change Indices, Bracknell, 1 Sep 1998, WMO TD 930

Frei C, Schär C (2001) Detection probability of trends in rare events: theory and application to heavy precipitations in the Alpine region. J Climate 14:1568–1584

IPCC (2007) Contribution of working group I to the fourth assessment report of the Intergovernmental Panel on Climate Change. In: Solomon S, Qin D, Manning M, Chen Z, Marquis M, Averyt KB, Tignorand M, Miller HL (eds). Cambridge University Press, Cambridge/New York

Keller F (2000) Les possibilités de changements dans la couverture végétale alpine peuvent-elle être condérées des risques, Actes du 5[e] Colloque transfrontalier CLUSE, Université de Genève

Messerli B, Ives JD (eds) (1997) Mountains of the world: a global priority. Parthenon, New York, p 495

Pepin NC, Seidel DJ (2005) A global comparison of surface and free-air temperatures at high elevations. J Geophys Res 110, D03104. doi:10.1029/2004JD005047

Weber RO, Talkner P, Stefanicki G (1994) Asymmetric diurnal temperature change in the Alpine region. Geophys Res Lett 21(8):673–676

Weber RO, Talkner P, Auer I, BoÈhm R, Gajic-Capka M, Zaninovic K, Bradzil R, Fasko P (1997) 20th century changes of temperatures in the mountain regions of Central Europe. Clim Change 36:327–344

# Chapter 10
# Projections of Future Changes in Climate of the Romanian Carpathians

**Abstract** This chapter briefly presents the changes of the air temperature and precipitation amounts predicted by Regional Climate Models for the next decades over the Romanian Carpathians. The analysis refers to the IPCC Scenario A1B, and exploits the outputs of several European projects developed in the recent years. The air temperature is likely to increase in all seasons, while the precipitation amounts will generally vary with ±10 % as compared to the present climate, at different spatial rates. The most significant temperature increasing is expected to occur in summer; over most Carpathian areas, the period 2021–2050 will be 2.5–3.0 °C warmer than 1961–1990. As regards the precipitation, the winter will be sensibly drier, while increasing trends are specific to the autumn.

While the characteristics of the present climate represent a topic of major interest, the humankind is highly concerned about the expected variability for the next decades. The Carpathian Mountains have a transnational influence over the environment and society, so that any climate changes may have a profound regional impact. The observed variability of the temperature and precipitation over the Carpathian region has been recently reported by Busuioc et al. (2010), based on ground station data, and by Cheval et al. (2014) and Spinoni et al. (2014), exploiting the outputs of the Carpatclim Project. In their turn, Birsan et al. (2014) tackled the variability of some extreme temperature and precipitation indices.

Busuioc et al. (2010) examined the connections between climate variability in Romania (air temperature and precipitation) and large-scale circulation. Accordingly, the increased frequency of high-level anticyclonic structures in winter, spring and summer, associated with a notable rise of mid-tropospheric temperature (850 hPa), may explain the ongoing warming, the occurrence of strong positive anomalies and high temperature extremes (particularly in summer), experienced in the last decades in many regions of this country, including the Carpathians. The large-scale atmospheric forcing controlling the observed changes in precipitation regime was found to be seasonally significant, in response to the increasing



D.M. Micu et al., *Climate of the Romanian Carpathians*,
Springer Atmospheric Sciences, DOI 10.1007/978-3-319-02886-6_10

frequency of cyclonic structures over northwestern Romania and the enhancement of zonal atmospheric circulations.

The results of ICT-RegCM climate simulations and downscaling models under the A2 IPCC scenario, for the period 2070–2099 relative to the 1961–1990 reference period summarized in the *Carpathians Environmental Outlook* (2007), indicate a general warming of about 5 °C in the Eastern Carpathians and 4.5–5.0 °C in the Southern and Western Carpathians (the Banat Mountains).

Busuioc et al. (2010) also investigated the climate projections over Romania, using extensively the results of the FP6-Project ENSEMBLES. Focussing on the Carpathian Mountains territory, we present the changes expected in the seasonal temperature and precipitation for the period 2021–2050, compared to the present climatology (1991–2020), and to the WMO reference period (1961–1990). The examination uses the A1B IPCC scenario, and the dataset is the resulted of the aggregation of three regional climate models (RCMs): RegCM3, ALADIN-climate, and PROMES (See Chap. 5.4 for more details).

In general, the variations between −5 % and 5 % in the precipitation amounts prevail, and deviations of ±10 % are common for most Carpathian areas, either comparing the future to the present or to the WMO reference, with specific particularities for each season (Figs. 10.1, 10.2, 10.3 and 10.4). A relative stability may be assumed for the spring, with slightly increasing trends over the inner Carpathian Chain, and spots of decreasing precipitation in the eastern and southern parts. One can expect less precipitation during the summer and especially during the winter seasons, while the autumn projections show a spatially extended increase

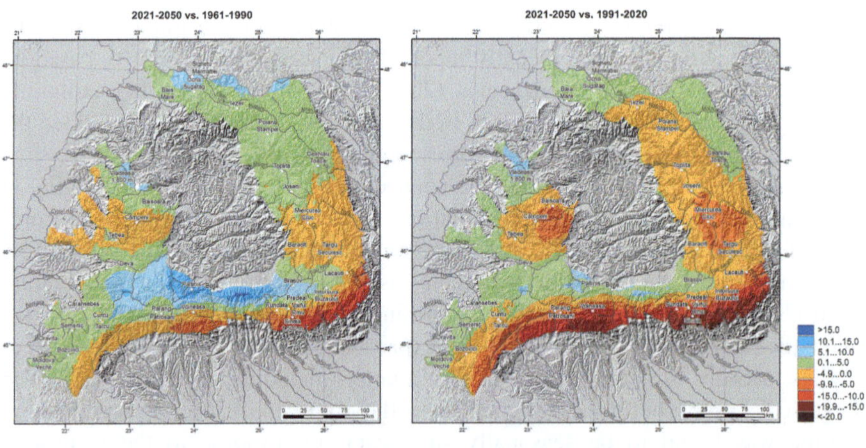


**Fig. 10.1** Spatial patterns of expected change (%) in winter precipitation in the Romanian Carpathians

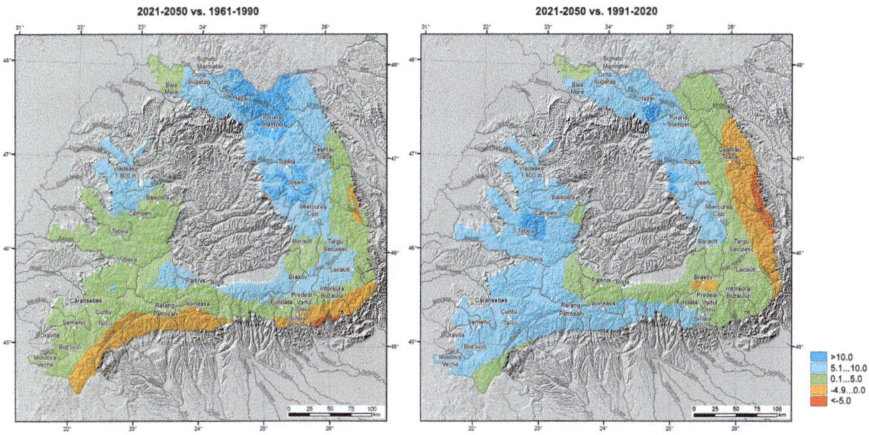


**Fig. 10.2** Spatial patterns of expected change (%) in spring precipitation in the Romanian Carpathians

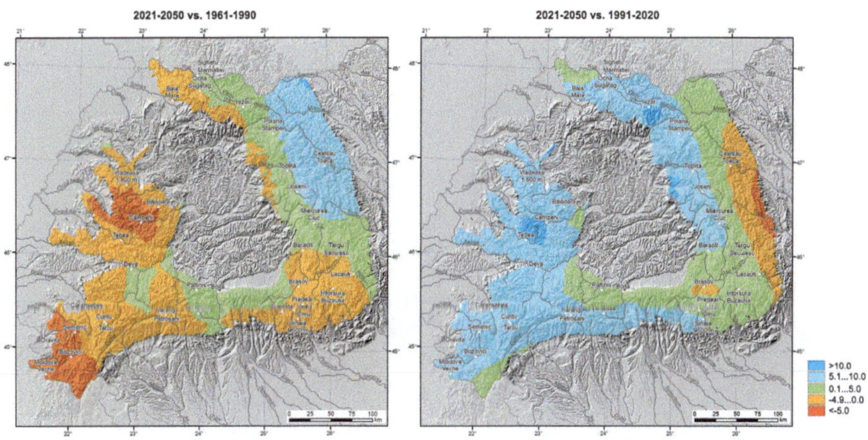


**Fig. 10.3** Spatial patterns of expected change (%) in summer precipitation in the Romanian Carpathians

over 2021–2050 compared to 1961–1990. However, the comparison to the current climate (1991–2020) reveals a decrease is likely in the western part of the Romanian Carpathians.

As regards the air temperature, the changing signal is more homogenous, largely positive over the entire region for all the seasons (Figs. 10.5, 10.6, 10.7 and 10.8).

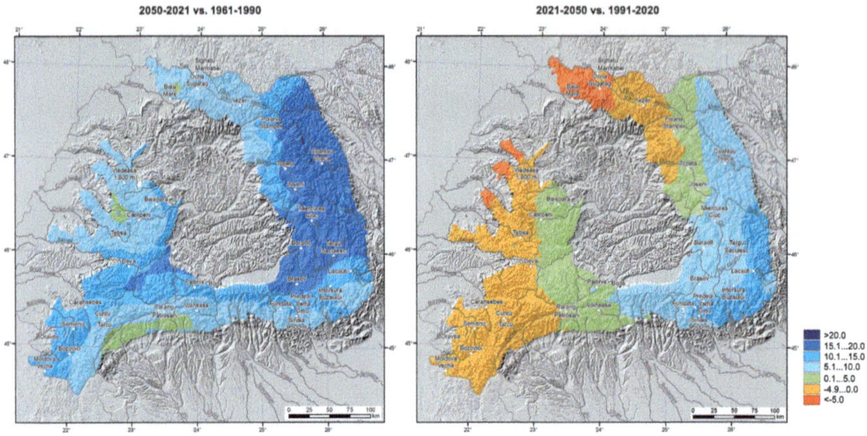


**Fig. 10.4** Spatial patterns of expected change (%) in autumn precipitation in the Romanian Carpathians

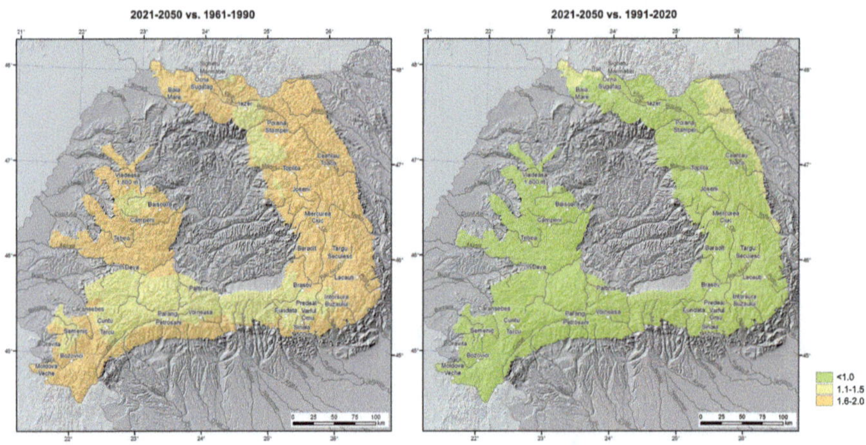


**Fig. 10.5** Spatial patterns of expected change (°C) in winter temperature in the Romanian Carpathians

The future will be dramatically warmer than the reference WMO period. The expected increase are likely to reach 1.6–2.0 °C in spring, autumn, and winter, and 2.5–3.0 °C in summer on most Carpathian areas. While compared to the present, the expected changes over the time interval 2021–2050 are still positive,

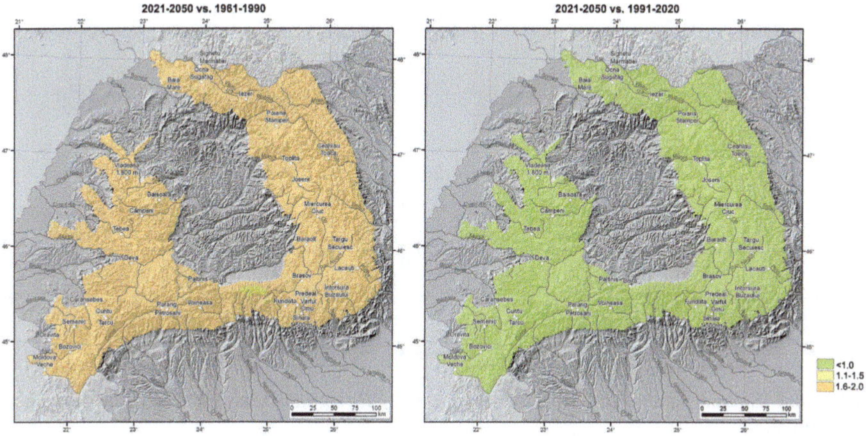


**Fig. 10.6** Spatial patterns of expected change (°C) in spring temperature in the Romanian Carpathians

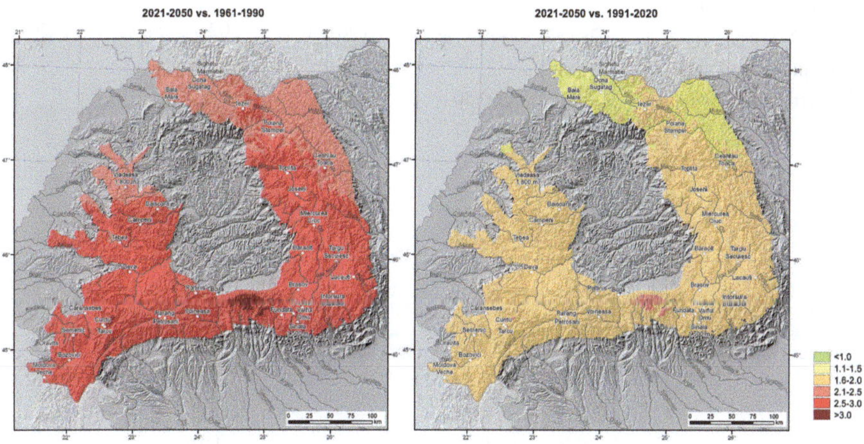


**Fig. 10.7** Spatial patterns of expected change (°C) in summer temperature in the Romanian Carpathians

but more temperate: 0.5–1.0 °C in spring, autumn, and winter, and 1.6–2.0 °C in summer.

The mean regional precipitation and temperature changes across the Romanian Carpathians are summarized in Table 10.1.

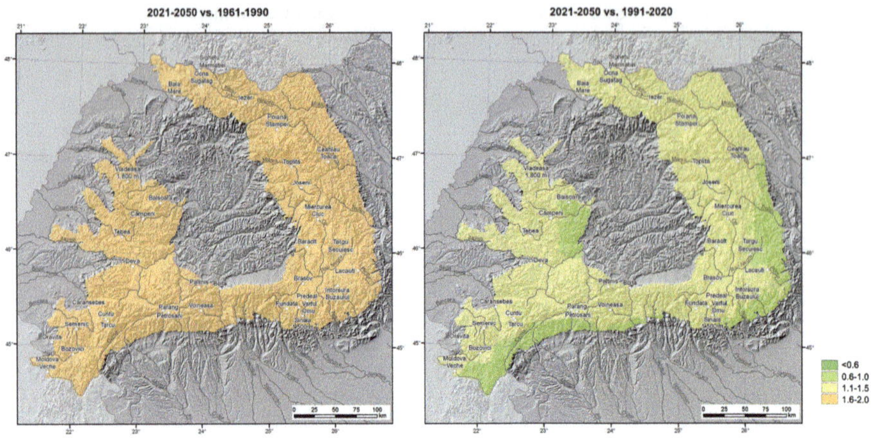


**Fig. 10.8** Spatial patterns of expected change (°C) in autumn temperature in the Romanian Carpathians

**Table 10.1** Projected mean regional precipitation (%) and mean temperature change (°C) in the Romanian Carpathians

| Romanian Carpathians regions | Season | 2021–2050 vs. 1961–1990 | 2021–2050 vs. 1991–2020 |
|---|---|---|---|
| Precipitation change | | | |
| Eastern Carpathians | DJF | 0.50 | −2.73 |
| | MAM | 6.37 | 3.43 |
| | JJA | 2.35 | 1.07 |
| | SON | 13.62 | 4.56 |
| Southern Carpathians | DJF | 3.10 | −4.64 |
| | MAM | 2.09 | 5.72 |
| | JJA | −0.63 | −2.65 |
| | SON | 9.45 | 2.80 |
| Western Carpathians | DJF | 1.17 | −0.56 |
| | MAM | 3.38 | 7.80 |
| | JJA | −4.27 | −2.93 |
| | SON | 9.32 | −1.16 |
| Temperature change | | | |
| Eastern Carpathians | DJF | 1.68 | 0.94 |
| | MAM | 1.76 | 0.81 |
| | JJA | 2.55 | 1.62 |
| | SON | 1.80 | 1.13 |
| Southern Carpathians | DJF | 1.58 | 0.87 |
| | MAM | 1.75 | 0.79 |
| | JJA | 2.75 | 1.78 |
| | SON | 1.79 | 1.11 |
| Western Carpathians | DJF | 1.63 | 0.90 |
| | MAM | 1.73 | 0.75 |
| | JJA | 2.59 | 1.65 |
| | SON | 1.77 | 1.12 |

# References


Birsan MV, Dumitrescu A, Micu DM, Cheval S (2014) Changes in annual temperature extremes in the Carpathians since AD 1961. Nat Hazards. doi:10.1007/s11069-014-1290-5

Busuioc A, Caian M, Cheval S, Bojariu R, Boroneanţ C, Baciu M, Dumitrescu A (2010) Climate variability and change in Romania (in Romanian). Edit. Pro Universitaria, Bucharest, 227 p

Cheval S, Birsan MV, Dumitrescu A (2014) Climate variability in the Carpathian Mountains Region over 1961–2010. Glob Planet Chang 118:85–96. doi:10.1016/j.gloplacha.2014.04.005

Spinoni J, Szalai S, Szentimrey T, Lakatos M, Bihari Z, Nagy A, Németh Á, Kovács T, Mihic D, Dacic M, Petrovic P, Kržič A, Hiebl J, Auer I, Milkovic J, Štepánek P, Zahradnícek P, Kilar P, Limanowka D, Pyrc R, Cheval S, Birsan MV, Dumitrescu A, Deak G, Matei M, Antolovic I, Nejedlík P, Štastný P, Kajaba P, Bochnícek O, Galo D, Mikulová K, Nabyvanets Y, Skrynyk O, Krakovska S, Gnatiuk N, Tolasz R, Antofie T, Vogt J (2014) Climate of the Carpathian Region in 1961–2010: climatologies and trends of ten variables. Int J Climatol. doi:10.1002/joc.4059

UNEP (2007) Carpathian environmental outlook – KEO2007. United Nations Environment Programme, 236 p. Available online at: unep.org/geo/pdfs/KEO2007_final_FULL_72dpi.pdf. Accessed Sept 2014

# Chapter 11
# Conclusions

'Complexity' is perhaps the term that defines mountain climatology in the most relevant manner, as it can be associated with many aspects in this direction. Mountain environments are always characterised by complex physical background generated at the interaction of many factors, leading to multiple consequences, creating specific resources, and impacting various social and economic interests. In this context, mountain climatology research has a crucial importance for understanding the relationships between the system's components, for preventing potential threats, and mitigating their impact, and for a sustainable use of the natural resources. This book is a climatic monograph which addresses one of the most complex mountain environments in Europe, the Carpathians Chain, focusing on the mountain regions which lie over Romania, herein called Eastern, Southern, and Western Carpathians.

To the best of our knowledge, based on the screening of the existing literature, this is the most comprehensive synthesis focusing on the climate of the Romanian Carpathians, providing sound scientific support for fundamental studies and applications, in the context of international initiatives and achievements related to the mountain environment. The current approach has been developed according to specific methodologies (i.e. homogenized data, long term periods, gridding techniques), but it also uses modern outputs (i.e. remote sensing products). It exploits the results of several international projects, but it is an independent work, with original deliverables meeting the needs of various end users.

The geographical background develops a close relationship with the Carpathian climate, defined by a temperate type, with less developed continental, Mediterranean or oceanic influences. The position of the mountain chain against regional atmospheric circulation, its elevation and morphology are the main terrestrial factors shaping the climate at various spatial and temporal scales. While the general characteristics remain dominant, there are still differences between the three major units of the Romanian Carpathians (Western, Southern, and Eastern), and local particularities also occur (i.e. glacial valleys, mountain corridors, intra-Carpathian depressions).


D.M. Micu et al., *Climate of the Romanian Carpathians*,
Springer Atmospheric Sciences, DOI 10.1007/978-3-319-02886-6_11

The core of the book contains a multifaceted analysis of the main climatic variables relevant for the area of interest: solar radiation, air temperature, precipitation, wind, and snow. Thorough statistical analysis, consistent efforts for spatialization (with a 1 km horizontal resolution), and investigations into the variability and trends over the period 1961–2010 were performed for each parameter.

Global and direct solar radiation fluxes have been investigated based on satellite remote sensing data over the period 1983–2005. The seasonal average values of the clear-sky global radiation vary between 50 and 60 W/m$^2$ in winter and 200–240 W/m$^2$ in summer.

Various aspects regarding the mean, maximum, and minimum air temperatures were discussed (lapse rates, vertical thermal zonation, annual regime, spatial distribution, and extreme temperature records), revealing both common, generalized characteristics, and particularities specific to different areas. For example, an overall regional lapse rate of about 0.60 °C/100 m was determined for the Romanian Carpathians based on the mean air temperature, with variations between the three regions. Altitudinal delineations were also found. On average, air temperature values become negative from November, at above 800 m and December, below 800 m, maintaining this characteristic until March, at above 800 m and February, below 800 m. Air temperature turns positive starting with April, in most areas of the Romanian Carpathians, except for high elevation areas above 1,800 m (since May) and earlier, from March or even February, generally below 600–700 m. The probability density functions (PDFs) of daily maximum temperatures show a limited heat stress exposure, while the daily minimum temperatures recorded above 1,000 m altitude reveal a moderate exposure to freezing.

As expected, the distribution of precipitation amounts is highly variable, and strongly dependent on the underlying topography and orographic bias to the regional flow dynamics. Statistically, altitude explains almost 60 % of the precipitation distribution in the Romanian Carpathians. The climate is rather moderately humid, as in more than 70 % of their surface area the annual precipitation is below 900 mm. Annual precipitation totals increase from 500 to 800 mm in the areas below 800 m, to 900–1,300 mm, at above 800 m. The Western and Southern Carpathians are the wettest, with a mean annual precipitation of about 800–900 mm, while the Eastern Carpathians are the driest (below 700 mm per year). There is an eastward decrease of precipitation amounts across the Romanian Carpathians, as the moisture content of the westerly and southwesterly (Mediterranean) airflows gradually diminish. The wind influence on the precipitation amounts has been presented in Sect. 7.2.3, and it helps the better understanding of the real water input to the mountain system. The 24-h precipitation records totaled 204.2 mm/July 19, 1970 at Cuntu (1,456 m) in areas above 800 m and 127.0 mm/September 1, 1941 at Caransebeş (241 m), in depressions.

The direction and strength of the winds across the Romanian Carpathians are widely controlled by the activity of the main high and low pressure systems operating at European scale: the Azores (all-year-round) and the East-European anticyclones (highly active during the winter months) and the Icelandic and the

Mediterranean cyclones (all-year-round). The westerly wind direction prevails in all high-elevation areas above 1,700–1,800 m, largely corresponding to the lower layers of zonal atmospheric flow. The elevation explains a statistically significant 52 % of the vertical zonation of wind speed within the Romanian Carpathians. The highest surface wind speed values occur in the Eastern and Southern Carpathians, above timberline, with annual averages exceeding 10 m/s in many sites. Spring is the windiest season, while the lowest speed values register during autumn.

Snow is an important component of the Carpathians landscape, and we present both snowfall and snowpack main characteristics, including general considerations about the triggering factors. The moist airflows and the troposphere vertical temperature profile depicting variations from 0 °C at ground level to −10 °C on cloudy peaks are important factors favouring the heavy snowfall in the area. The snow occurrence highly depends on the synoptic-scale influences and on topographic conditions, i.e. altitude, slope aspect. Usually, the first snowfalls appear from early September in areas above 2,000 m, to early-to-mid November, in depressions, while the last winter snowfalls are likely from late April to early July, in the areas above 800 m, and from late March to mid-April, in depressions. The snowfall events occur with different frequencies and intensities, with large spatial and temporal variations. Generally, in the Romanian Carpathians, the snow falls in 40–80 days during the extended winter season (November–April), with higher values in the areas above 1,700–1,800 m, and lower values specific for depressions and areas below 1,000 m (25–40 days/season). The increase of cyclonic activity during the late winter to early spring interval, when snowfall events are frequently abundant, explains the occurrence of snow depth peak over the February–March interval in most areas. The maximum snow depth records over the 1961–2010 period ranged between 109 and 145 cm in the areas above 800 m and 30–42 cm in those below 800 m.

Consistent efforts have been developed for investigating the variability and trends of the main climatic elements during the period 1961–2010, with a special section dedicated to the variability of the extremes. Temperature trends are increasing in winter, spring and summer, while they are absent in autumn. On the other hand, autumn is the only season when significant increasing trends in precipitation have been found. The average wind speed is decreasing in all seasons. The analysis of the snow variability shows general decreasing trends in mean snow depth, number of days with snow cover, and number of days with snowfall. The decreasing trend in the maximum length of snowfall spells is related to recent warming than to precipitation decrease.

The climatic variability for the next decades has been tackled for the A1B IPCC scenario, which considers a good balance between all sources of energy, and a rapid economic development by the mid-twenty first century. This work exploits the results produced within several European projects, and the findings are consistent with previous publications. Comparing the climatology over the periods 2021–2050 and 1991–2020, the air temperature are very likely to increase with different rates, while the seasonal precipitation amounts will probably vary between ±10 %.

*Climate of the Romanian Carpathians. Variability and Trends* has the ambition to be the most comprehensive and detailed climatic monograph over the area, aggregating high quality input data, up-to-date techniques, regional analysis, and overview perspectives, while addressing the spatial and temporal patterns of the main climate elements. Nevertheless, we are confident that future investigations may bring even more useful knowledge about the Romanian Carpathians, and we warmly encourage any criticism and follow-up.

# Index




D.M. Micu et al., *Climate of the Romanian Carpathians*,
Springer Atmospheric Sciences, DOI 10.1007/978-3-319-02886-6

MIX
Papier aus verantwortungsvollen Quellen
Paper from responsible sources
FSC® C105338

If you have any concerns about our products,
you can contact us on
**ProductSafety@springernature.com**

In case Publisher is established outside the EU,
the EU authorized representative is:
**Springer Nature Customer Service Center GmbH**
**Europaplatz 3, 69115 Heidelberg, Germany**

Printed by Libri Plureos GmbH
in Hamburg, Germany